SpringerBriefs in Applied Sciences and Technology

PoliMI SpringerBriefs

Series Editors

Barbara Pernici, DEIB, Politecnico di Milano, Milan, Italy
Stefano Della Torre, DABC, Politecnico di Milano, Milan, Italy
Bianca M. Colosimo, DMEC, Politecnico di Milano, Milan, Italy
Tiziano Faravelli, DCHEM, Politecnico di Milano, Milan, Italy
Roberto Paolucci, DICA, Politecnico di Milano, Milan, Italy
Silvia Piardi, Design, Politecnico di Milano, Milan, Italy
Gabriele Pasqui , DASTU, Politecnico di Milano, Milan, Italy

Springer, in cooperation with Politecnico di Milano, publishes the PoliMI Springer-Briefs, concise summaries of cutting-edge research and practical applications across a wide spectrum of fields. Featuring compact volumes of 50 to 125 (150 as a maximum) pages, the series covers a range of contents from professional to academic in the following research areas carried out at Politecnico:

- Aerospace Engineering
- Bioengineering
- Electrical Engineering
- Energy and Nuclear Science and Technology
- Environmental and Infrastructure Engineering
- Industrial Chemistry and Chemical Engineering
- Information Technology
- Management, Economics and Industrial Engineering
- Materials Engineering
- Mathematical Models and Methods in Engineering
- Mechanical Engineering
- Structural Seismic and Geotechnical Engineering
- Built Environment and Construction Engineering
- Physics
- Design and Technologies
- Urban Planning, Design, and Policy

Mario Grosso · Lucia Rigamonti
Editors

Waste Flows Generated by the Energy Transition

Regulatory Framework, Recovery Technologies and Plant Infrastructures

Editors
Mario Grosso
Department of Civil and Environmental
Engineering, DICA
Politecnico di Milano
Milan, Italy

MatER Study Center
Laboratory for Energy and the Environment
Piacenza
Piacenza, Italy

Lucia Rigamonti
Department of Civil and Environmental
Engineering, DICA
Politecnico di Milano
Milan, Italy

MatER Study Center
Laboratory for Energy and the Environment
Piacenza
Piacenza, Italy

ISSN 2191-530X   ISSN 2191-5318 (electronic)
SpringerBriefs in Applied Sciences and Technology
ISSN 2282-2577   ISSN 2282-2585 (electronic)
PoliMI SpringerBriefs
ISBN 978-3-031-88950-9   ISBN 978-3-031-88951-6 (eBook)
https://doi.org/10.1007/978-3-031-88951-6

© The Editor(s) (if applicable) and The Author(s), under exclusive license to Springer Nature Switzerland AG 2025

This work is subject to copyright. All rights are solely and exclusively licensed by the Publisher, whether the whole or part of the material is concerned, specifically the rights of translation, reprinting, reuse of illustrations, recitation, broadcasting, reproduction on microfilms or in any other physical way, and transmission or information storage and retrieval, electronic adaptation, computer software, or by similar or dissimilar methodology now known or hereafter developed.
The use of general descriptive names, registered names, trademarks, service marks, etc. in this publication does not imply, even in the absence of a specific statement, that such names are exempt from the relevant protective laws and regulations and therefore free for general use.
The publisher, the authors and the editors are safe to assume that the advice and information in this book are believed to be true and accurate at the date of publication. Neither the publisher nor the authors or the editors give a warranty, expressed or implied, with respect to the material contained herein or for any errors or omissions that may have been made. The publisher remains neutral with regard to jurisdictional claims in published maps and institutional affiliations.

This Springer imprint is published by the registered company Springer Nature Switzerland AG
The registered company address is: Gewerbestrasse 11, 6330 Cham, Switzerland

If disposing of this product, please recycle the paper.

Preface

The term "energy transition" refers to the global shift from reliance on fossil fuels to renewable energy sources such as solar, wind, hydro, biomass, and geothermal energy. This transition also involves advancements in energy storage technologies, widespread electrification of end-uses, improved energy efficiency, and enhanced demand management strategies. It is driven by pressing concerns over climate change, air pollution, energy security, and socio-economic sustainability. Among these, it is the climate change concern that is paving the way, with most countries committed to reducing fossil fuel emissions to meet the Paris Agreement's ambitious goal of limiting global temperature increases to well below 2°C—and ideally to 1.5°C—above pre-industrial levels.

In the coming decades, the energy sector is poised for a profound transformation, moving from a fossil-fuel-dominated landscape (coal, oil, and natural gas) to one centred on renewable energy and cleaner technologies. The remarkable growth of solar and wind power, the adoption of electric vehicles, and other technological advancements illustrate the urgent and widely acknowledged need to mitigate greenhouse gas emissions and reduce the environmental impacts of a carbon-intensive economy.

However, the shift to cleaner energy technologies comes with a significant demand for critical minerals and metals. For instance, lithium, cobalt, manganese, phosphorus, and nickel are essential for batteries; rare earth elements are critical for permanent magnets in wind turbines and electric motors; nickel and platinum group metals are used in hydrogen electrolysers and fuel cells; and copper and aluminum are indispensable for electricity-related technologies and distribution networks. The uneven geographic distribution of these materials and bottlenecks in their supply chains pose challenges to the expansion of renewable energy and electricity-based technologies.

From a materials perspective, the energy transition signifies a shift from a linear model—centred on single-use fossil fuels—to a circular model. Here, energy is sourced freely, while the materials required for production, transformation, and transportation infrastructures are reused, repurposed, and recycled as much as possible.

The reliance on critical minerals in the energy transition necessitates thorough investigation to ensure climate change mitigation efforts do not inadvertently cause adverse effects, such as burden-shifting to other environmental impact categories. Moreover, the lifecycle of clean technologies inevitably generates waste, including end-of-life products, mining byproducts, processing tailings, and manufacturing scraps.

Effective management of these waste streams is vital. On one hand, improper disposal must be avoided to prevent environmental harm; on the other, recycling emerges as a key strategy to meet growing material demands. Recycling reduces dependency on virgin raw materials, alleviates supply chain pressures, and enhances resource security. However, successful recycling requires not only technological solutions but also efficient systems for the collection, separation, and processing of waste materials.

This book explores four major waste streams associated with the energy transition: (i) lithium-ion batteries for electric vehicles and stationary energy storage, (ii) electric motors in electric vehicles, (iii) wind turbine blades, and (iv) photovoltaic panels. For each technology, the current options for end-of-life (EoL) management are analysed, including a detailed characterization of the components and materials in the corresponding waste stream. The legislative framework is examined to evaluate whether it fosters or hinders circular economy initiatives, and industrial practices and research efforts in repurposing and recycling are summarized.

We aspire for this book to serve as a resource for scholars, students, and practitioners in the fields of energy transition and sustainability. Our goal is to address the concerns and challenges surrounding the phase-out of fossil fuels, ensuring they are tackled comprehensively and responsibly.

The research underpinning this book was conducted by the MatER (Materials and Energy from Refuse) research centre, established in 2011 in Piacenza, Italy, within LEAP (Laboratorio Energia e Ambiente Piacenza), an external laboratory of Politecnico di Milano. MatER's mission is to identify efficient, sustainable waste management solutions and assess the best available techniques for material and energy recovery, exploring synergies between these two pillars of recovery from waste.

We would also like to thank the master's students of Politecnico di Milano who supported the initial phases of the research on the state-of-the-art: S. Bruschi, L. Ferrari, S. Ladisa, P. Salardi, I. Scigliuolo, and D. Varotto for their work on Li-ion batteries, and J. D. Chamat Torres for their contributions on waste from wind turbines.

For the field visits and exploratory calls, we extend our gratitude to P. Rädecker from Batrec AG, D. Baldi from Riraee Srl, V. Ramon and L. Ramon from Tialpi Srl, F. Miserocchi from 9-Tech, G. Perben and P. Gallo from Composite Recycling, G. Bonati from Rivierasca SpA, Professor M. Colledani as the lead for the FibeReuse and DEREMCO projects, and D. Grisenti from Rina Consulting, the project manager of the REFRESH project.

Milan, Italy

Mario Grosso
Lucia Rigamonti

About this Book

This book analyses four critical waste streams linked to the transition to cleaner energy technologies: Li-ion batteries for electric vehicles and energy storage systems, electric vehicle motors, composite waste from wind turbines, and photovoltaic panels.

Each chapter explores waste stream characteristics, including recoverable critical raw materials, legislative frameworks, waste prevention and management options, and surveys on industrial initiatives and facilities in Europe.

The text examines the state-of-art of recovery technologies for extracting valuable resources to meet the rising demand for critical materials. It reviews reuse, remanufacturing, and recycling methods, including mechanical, hydrometallurgical, pyrometallurgical, and chemical recovery processes. Each technology is discussed with its respective key features, strengths and drawbacks, also from the environmental standpoint.

A thorough evaluation of European regulatory frameworks for end-of-life management is provided.

The book also evaluates existing plant infrastructure across Europe, identifying potential gaps in facilities that may challenge waste treatment and recycling in the future.

This resource is aimed at policymakers, industry stakeholders, researchers, students, and practitioners focused on sustainable waste management in the energy transition.

Contents

Introduction: Recycling to Support the Energy Transition 1
Gaia Brussa and Stefano Puricelli

Lithium-Ion Batteries 13
Stefano Puricelli

Electric Motors 45
Stefano Puricelli and Sebastián Fajardo Turner

Wind Turbine Blades 73
Gaia Brussa

Photovoltaic Panels 103
Gaia Brussa

Introduction: Recycling to Support the Energy Transition

Gaia Brussa and Stefano Puricelli

Abstract The energy transition represents a global shift from fossil fuels to renewable energy sources such as solar, wind, and hydropower, driven by the urgent need to address climate change, air pollution, and energy security. This transition, however, intensifies the demand for critical and strategic raw materials required for clean energy technologies, including lithium, cobalt, rare earth elements, and copper. These materials are essential for manufacturing and maintaining technologies like batteries, wind turbines, photovoltaic panels, and electric motors. The European Union (EU) faces significant supply chain vulnerabilities, as the extraction and processing of these materials are heavily concentrated in non-EU countries, such as China. Recycling emerges as a pivotal strategy to mitigate these risks by reducing dependence on virgin resources, enhancing material recovery, and addressing waste streams from clean energy technologies. Current recycling processes effectively handle bulk metals like steel and copper but struggle with emerging waste streams, such as lithium-ion batteries and rare earth magnets, due to technical and economic challenges. This introduction explores the material requirements for four crucial technologies—batteries, electric motors, wind turbine blades, and photovoltaic panels—and evaluates the potential of recycling to secure supply chains and minimise environmental impacts. By analysing trends and policies, such as the EU's Critical Raw Materials Act, the introduction underscores the importance of achieving ambitious recycling targets and fostering innovation in waste management technologies to support a sustainable energy transition.

Keywords Energy transition · Renewable energy · Critical Raw Materials · Waste · Recycling

---

G. Brussa (✉) · S. Puricelli
Department of Civil and Environmental Engineering, Politecnico di Milano, Milano, Italy
e-mail: gaia.brussa@polimi.it

MatER Study Center, Laboratory for Energy and the Environment Piacenza, Piacenza, Italy

S. Puricelli
e-mail: stefano.puricelli@polimi.it

© The Author(s), under exclusive license to Springer Nature Switzerland AG 2025
M. Grosso and L. Rigamonti (eds.), *Waste Flows Generated by the Energy Transition*, PoliMI SpringerBriefs, https://doi.org/10.1007/978-3-031-88951-6_1

1 The Context of Energy Transition

The term "energy transition" refers to the global tendency in the energy sector to move away from the reliance on fossil fuels and shift to renewable sources (such as solar, wind, hydro, biomass, and geothermal energy) coupled with energy storage technologies, as well as improve energy efficiency and demand management strategies. This transition is driven by concerns over climate change, air pollution, energy security, as well as economic sustainability.

In the coming decades the energy sector is expected to face a profound transformation from the one dominated by the use of fossil fuels (such as coal, oil, and natural gas) to one increasingly based on renewable energy sources and cleaner technologies: a rapid growth of solar and wind power, as well as the increase of energy storage capacity and electric vehicles penetration, together with a range of other technologies, is expected in order to tackle the now widely accepted need to reduce greenhouse gases emissions and the environmental impacts (i.e., air pollution) of a carbon-intensive energy.

In the last decade, the supply of energy from renewable sources, replacing fossil-based ones, has accelerated, and the uptake of clean energy technologies is expected to continue growing. In 2022, renewables reached 30% of the global power generation, compared to 20% in 2010, thanks to solar photovoltaic (PV), wind, hydropower, and bioenergy (IEA, 2023b). This growth will be fostered also by electrification that is accelerating across all end-use sectors (IEA, 2023a).

1.1 Materials for Clean Energy Technologies

The energy transition to cleaner technologies is expected to be material-intensive and to drive a major increase in the demand of minerals and metals. Contrarily to fossil-based technologies, whose impacts mainly come from the utilisation phase, the potential environmental issues related to renewable energies are linked to the materials used for their manufacturing and during maintenance (IEA, 2021b).

Especially metals play a pivotal role in all technologies required for the energy transition; they can be divided into three groups (Gregoir & van Acker, 2022):

- base metals such as aluminium, copper, nickel and zinc: they have a wide range of uses and are strategic to the production of most technologies;
- metals with specific uses: silicon and tellurium for photovoltaic cells, rare earth elements for permanent magnets used in electric motors and generators, lithium, cobalt, nickel, manganese, phosphorous for batteries;
- alloying elements (e.g., vanadium, molybdenum, manganese): commonly used in steel that is largely employed in various technologies, although it is not easy to assess which alloyed steel is required for each application.

One particularly relevant category of metals is the group of rare earth elements (REEs). REEs are a group of 17 chemicals containing the lanthanides (15 elements), yttrium and scandium, as the latters are always found with the other metals. REEs are split into two groups: the Light Rare Earth Elements (LREE)[1] and the Heavy Rare Earth Elements (HREE)[2] both for physic-chemical and commercial reasons (European Commission, 2020). REEs are used in various high-tech applications such as smartphones, hard disk drives, LEDs lights, but also to produce permanent magnets, used both for electric motors and wind turbines generators; especially dysprosium (Dy), neodymium (Nd), terbium (Tb) and praseodymium (Pr) are needed in these applications (Gregoir & van Acker, 2022).

The European Union (EU) is planning to strongly improve the generation and use of renewable energy. A huge increase in clean energy technologies will therefore increase the need for raw materials which are largely controlled by non-EU countries. In fact, many of the required resources are unevenly distributed and the geographical distribution is more concentrated compared to oil or natural gas, and their supply chains are constrained by bottlenecks since they are produced or processed in a handful of countries, posing a threat to renewables' expansion (Carrara et al., 2020). Lithium is extracted primarily in Australia and Chile, cobalt in the Democratic Republic of Congo (DRC), and nickel in Indonesia; the main sources of copper are in South America (Chile and Peru) and more than 60% of the global production of REEs is concentrated in China (IEA, 2021b). Furthermore, the processing and refining of ores into minerals is even more centralised: in fact, this phase is carried out mainly in China for all the abovementioned materials. In 2019, in terms of processing volumes, China controlled from about 35% (nickel) to more than 90% (REEs) of the activities, while the usage was concentrated in the EU and the US (IEA, 2021b).

In this context, raw materials which are considered crucial for the EU economy have been added over time to the list of the so-called Critical Raw Materials (CRMs). This list is subject to change depending on the evolving factors for each of the selected materials and the latest version, which was published in 2023, includes 34 raw materials. The assessment of criticality depends on both the importance for the EU economy (Economic Importance—EI) and the risk associated with the supply (Supply Risk—SR). EI is based on the economic market value, while SR relates to the technical properties, the extraction difficulties, the strength of the supply chain, the political stability and the policies of the countries in which extraction is performed, as well as the social impacts and regulations that are connected to such materials. In Grohol and Veeh (2023) the fifth assessment of CRMs is presented: the methodology for the definition of a critical raw material by European standards is applied to 87 candidate raw materials and results supporting the list of 34 CRMs are reported. In addition, starting from 2023, a subgroup of 16 CRMs has been identified as Strategic

---

[1] LREEs include *cerium (Ce), lanthanum (La), neodymium (Nd), promethium (Pm), praseodymium (Pr), samarium* (Sm).
[2] HREEs include *dysprosium (Dy), erbium (Er), europium (Eu), gadolinium (Gd), holmium (Ho), lutetium (Lu), terbium (Tb), thulium (Tm), ytterbium (Yb), yttrium (Y)*.

Raw Materials i.e., a group of materials linked to a number of industrial sectors and technologies. Among others, strategic raw materials include copper and nickel, which do not meet the thresholds for CRMs assessment but are regardless included in the CRMs list (European Commission, 2023a). This addition was supported by the intensive use of the two metals, mainly due to electrification in the case of copper, which is hardly replaceable, and due to the employment in batteries and the increasing concentration of ownership of extraction and production sites in case of nickel.

Critical and strategic raw material lists have been identified by the Critical Raw Materials Act, a regulation proposed by the European Commission to reduce the EU dependency from other countries and secure the needed raw materials (European Commission, 2023a). Briefly, the proposal sets the following targets for the EU's annual consumption of strategic raw materials by 2030: (i) to extract at least 10%; (ii) to process at least 40%; (iii) to recycle at least 15%; (iv) to not rely for more than 65% of each strategic raw material from a single third country, at any relevant stage of processing.

Table 1 reports the latest list of critical and strategic raw materials as included in the Critical Raw Material Act.

It has been observed that, generally, renewable energy generation and storage as well as electric mobility require more materials than their conventional alternatives: not only greater amounts are required (as shown in the comparisons in Fig. 1), but also the demand for critical and rare materials is higher than for fossil-based technologies.

Nevertheless, the types of mineral resources vary depending on the specific technology (IEA, 2021b):

- lithium, cobalt, and nickel play a central role in batteries i.e. for better performance, longevity and higher energy density;
- rare earth elements are used for permanent magnets necessary in wind turbines and electric motors;
- nickel or platinum group metals are required by hydrogen electrolysers and fuel cells, depending on the type of technology;
- copper, together with aluminium, is an essential element for almost all electricity-related technologies and networks.

A study by Carrara et al. (2023) highlights the connection between energy transition and raw material demand, in particular minerals and metals that clean energy systems require in higher amount with respect to the fossil-fuels energetic system. The report (Carrara et al., 2023) identifies a group of strategic technologies and their main materials, components and assemblies. Also, material demand forecast scenarios are defined, and the related supply chains and possible bottlenecks are explored. The 15 strategic technologies are lithium-ion batteries, fuel cells, electrolysers, wind turbines, traction motors, solar photovoltaics, heat pumps, hydrogen direct reduced iron and electric arc furnaces, data transmission networks, data storage and servers, smartphones, tablets and laptops, additive manufacturing, robotics, drones, space launchers and satellites. The study considers 87 raw materials which can be divided into strategic, critical and non-critical.

Introduction: Recycling to Support the Energy Transition

Table 1 List of Critical and Strategic Raw Materials according to the Proposal of Critical Raw Material Act, 2023 (European Commission, 2023a). Copper and nickel are SRMs also included in the list of CRMs, although not meeting the CRM thresholds

Critical Raw Materials (CRMs)	Strategic Raw Materials
Antimony	
Arsenic	
Bauxite	
Baryte	
Beryllium	
Bismuth	Bismuth
Boron	Boron—metallurgy grade
Cobalt	Cobalt
Coking coal	
Copper	Copper
Feldspar	
Fluorspar	
Gallium	Gallium
Germanium	Germanium
Hafnium	
Helium	
Heavy rare earth elements	Rare earth elements for magnets: Ce, Sm, Tb,
Light rare earth elements	Rare earth elements for magnets: Dy, Gd, Nd, Pr
Lithium	Lithium—battery grade
Magnesium	Magnesium metal
Manganese	Manganese—battery grade
Natural graphite	Natural graphite—battery grade
Nickel—battery grade	Nickel—battery grade
Niobium	
Phosphate rock	
Phosphorus	
Platinum group metals (PGMs): Ir, Pd, Pt, Rh, Ru	Platinum group metals (PGMs): Ir, Pd, Pt, Rh, Ru
Scandium	
Silicon metal	Silicon metal
Strontium	
Tantalum	
Titanium metal	Titanium metal
Tungsten	Tungsten
Vanadium	

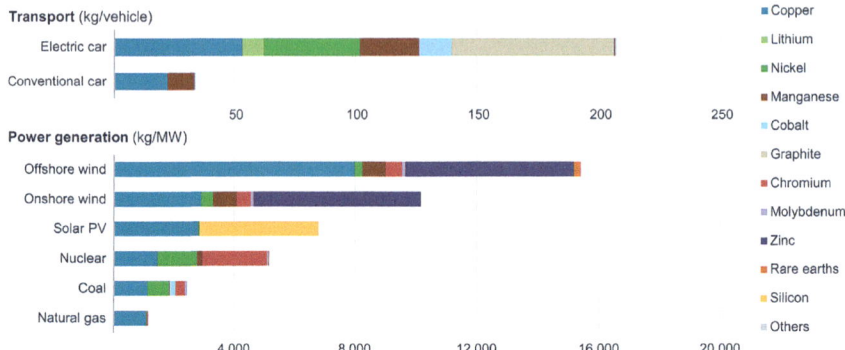

Fig. 1 Raw material requirements for different car and energy technologies. *Note* The values for electric car are based on a 75 kWh NMC (nickel manganese cobalt) 622 cathode and graphite-based anode. The values for offshore wind are based on direct-drive permanent magnet synchronous generator (including array cables) and for onshore wind on doubly fed induction generator system. The values for coal are based on ultra-supercritical plants and for natural gas on combined-cycle gas turbines. Reprinted from (IEA, 2021b); license: CC BY 4.0

1.2 The Role of Recycling

The reliance of energy transition on critical minerals (copper, lithium, aluminium, rare earth metals, cobalt, etc.) (Gregoir & van Acker, 2022) is a challenge that must be investigated to fully understand their potential in fighting climate change without disregarding other possible impact categories, i.e., avoiding burden shifting, and accurately assess the carbon offsets provided. Additionally, the fact that all these clean technologies are going to generate waste during their life cycle must be properly considered. On the one hand, there is the need to address the expected waste streams and properly manage them, preventing environmental impacts of improper disposal. On the other hand, there is the increasing demand for materials, which put pressure on the supply of virgin raw materials. Recycling emerges as a key strategy to face both challenges, helping to manage waste effectively while reducing reliance on new resources and improving the security of their supply, thereby mitigating dependency on other countries.

Waste related to clean energy technologies comprises the end-of-life products, but also waste from mining, tailing from processing and scraps from manufacturing. Managing this waste can be particularly relevant for critical minerals and metals. Improved recovery of metals from mining and processing byproducts (such as mining residues, slag, sludges, and tailings) presents the opportunity for improved supply. Also, enhanced waste stream management mitigates the potential for hazardous materials to leach into the environment (IEA, 2021b).

Recycling requires not only the technology to process the waste but also the physical collection and separation of waste streams and materials to be treated. The recycling processes for bulk metals (e.g., iron, steel, copper, aluminium) are well established but this is not the case for other metals required for the energy transition, e.g., lithium and REEs (IEA, 2021b). The increase in emerging waste streams such as wind turbines and batteries can give a boost to investments in technologies to treat that waste, especially in regions with a wider deployment of these clean energy technologies thanks to economies of scale (IEA, 2021b).

According to the IEA (2021b), the level of recycling can be measured by two indicators:

- End-of-life recycling rate (EOL-RR);
- Recycling input rates, also called recycled content rates (EOL-RIR).

The EOL-RR measures the share of material in post-consumer waste flows that results as secondary material from the recycling process with respect to the generated waste (European Commission & EIP on Raw Materials, 2021); it is an output-related indicator and the value results from the multiplication of the collection rate and the recycling efficiency (Binnemans et al., 2013). Instead, the EOL-RIR aims at assessing the share of secondary sources covering the material demand i.e., the recycled material inputs with respect to the total material input to the production system (IEA, 2021b).

EOL-RIR indicator may be difficult to interpret since it is determined by two different and independent factors: raw material demand and amount of recyclable material, i.e., material in waste which is available for recycling. Indeed, the EU can recycle significant quantities of scrap but may still show decreasing EOL-RIR values due to the tendency of increasing demand for certain materials; especially this has been observed for several CRMs in some low-carbon or digital technologies. The opposite trend can be observed in case of materials bans that drastically reduce the demand. For bulk metals that are already widely and efficiently recycled, the EOL-RIR remains still relatively low primarily because they are incorporated into long-life capital goods which will become available for recycling only in the future. For many materials, including numerous CRMs, the EOL-RIR is negligible or very low since they have been recently introduced in innovative and complex products (such as electric vehicles, renewable energy plants, and electronics), and recycling technologies for these materials are either not yet available or not profitable (European Commission, 2023b).

Many initiatives have been carried out in the EU on the topic of raw materials; for example, starting from the Circular Economy Action Plan (European Commission, 2015), the JRC has developed and periodically updates a Raw Materials Information System (RMIS) with the aim of providing knowledge on raw materials, mainly metals and minerals, and the associated value chains, including supply risks and circularity. The updated version on the interactive online tool was released in 2023 (European Commission & Joint Research Centre, n.d.). As part of the RMIS, a Raw Materials Scoreboard is published every two years in order to give monitoring information, organised in 27 indicators, on raw materials. Indeed, one indicator is the EOL-RIR, since it reflects not only the role of recycling in circular economy but also its importance to reduce the dependency on imports and to improve security of supply (European Commission & Joint Research Centre, n.d.).

Although recycling rates (EOL-RR) show promising values, up to above 60%, the contribution of recycled materials to the final demand (EOL-RIR) is generally still much lower (Fig. 2).

The highest EOL-RIR is reported for lead (75%) followed by metals for which the collection and recycling routes are well-established such as iron ore (31%), tin (31%) and zinc (31%). Considering CRMs, interesting EOL-RIRs are showed by yttrium (31%), palladium (28%), rhodium (28%) and platinum (25%), while for major metals (e.g., aluminium, chromium, copper, magnesium, and nickel) and precious metals, as silver, the values are in the range 10–25%. Regarding the materials employed in

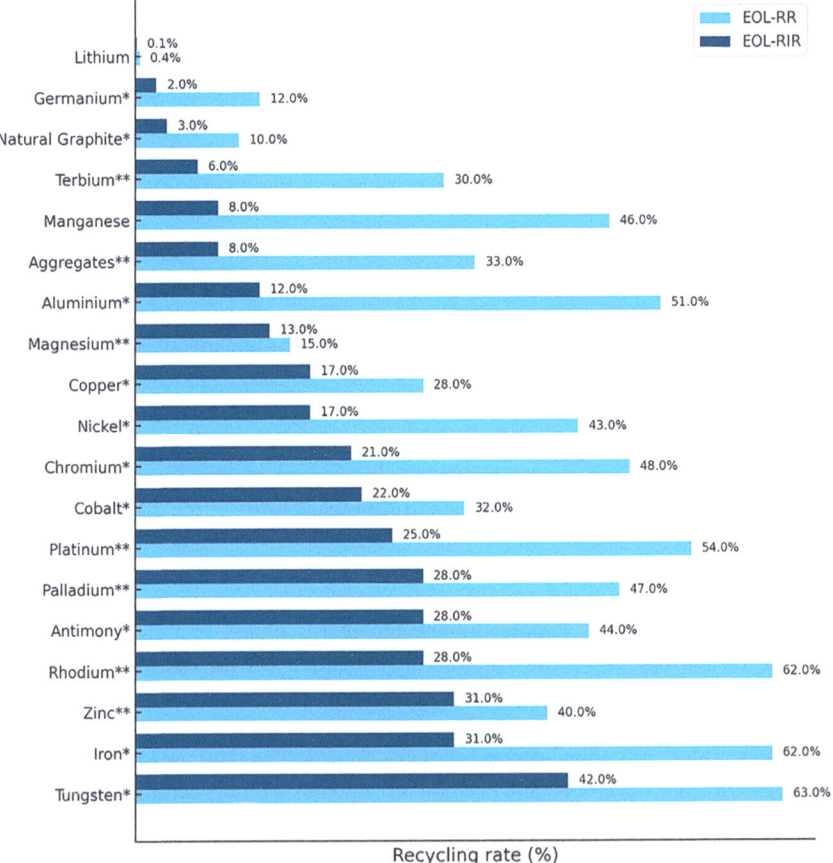

Fig. 2 End-of-Life recycling rates (EOL-RR) and end-of-life recycling input rates (EOL-RIR) for a selection of materials. Note that recycling estimations are based on: *EC Material System Analysis Studies. Geographical coverage: EU **European Commission (2020) (EOL-RIR) and UNEP (2011) (EOL-RR). Geographical coverage: EU/Global. Reproduced from European Commission and EIP on Raw Materials (2021)

batteries, cobalt presents a relatively relevant EOL-RIR (22%) while lithium is still at minimum levels of EOL-RIR and EOL-RR, just slightly above 0%.

However, it should be noted that also recycling rates (EOL-RR) show a significant variability and depend greatly on the metals in question. The efficiency has been improving in recent years for common metals and materials (copper, aluminium, nickel, iron), but new products and complex alloys are a significant challenge for the end-of-life stage, since their treatment is still expensive and needs dedicated technologies on a case-by-case basis (IEA, 2021b).

1.3 Waste Flows Generated by the Energy Transition

This book explores four clean energy technologies and related end of life: lithium-ion batteries (LIBs) for electric vehicles and stationary energy storage, electric motors for electric vehicles, composite materials for wind turbine blades, and photovoltaic panels.

The rationale behind the selection of the four waste streams can be understood by considering the future projections on the energy transition. For example, in the special report "Net Zero by 2050: A roadmap for the global energy system" (IEA, 2021a), the International Energy Agency tries to provide a comprehensive analysis of the necessary transformation of the global economy and the energy system to reach the goal of net zero greenhouse gas (GHG) emissions by 2050. The starting point of the report is global energy projections based on the Stated Policies Scenario (STEPS), which includes current and announced policies, such as Nationally Determined Contributions (NDCs). Then, it explores the Announced Pledges Case (APC), assuming full implementation of all announced net zero targets (Fig. 3) and highlights the required changes across sectors and technologies such as PV, wind, batteries and electric vehicles (EV) are extensively considered. In fact, renewable energy sources will play a pivotal role in the decarbonisation of electricity generation: the share of renewables in electricity generation increases from 29% in 2020 to nearly 70% by 2050, driven primarily by the rapid growth of solar PV and wind, which together will supply almost half of the electricity by mid-century (IEA, 2021a).

Moreover, the report presents the Net-Zero Emissions by 2050 Scenario (NZE), outlining how energy demand and the energy mix must evolve to reach net-zero emissions by 2050: solar PV and wind energy, paired with battery storage, are expected to be the key in this transition. To be in line with the net zero emission (NZE) pathway, worldwide, solar PV installed capacity needs to be growing 20-fold and wind energy 11-fold by 2050. Moreover, annual additions need to be four-times the levels of 2020, reaching 630 GW of solar and 390 GW of wind by 2030 (IEA, 2021a).

For the transport sector, instead, the technological transition relies on electric mobility, especially battery electric vehicles (BEVs), plug-in hybrid electric vehicles (PHEVs) and fuel cell electric vehicles (FCEVs), allowing the shift of a sector which is still heavily relying on fossil fuels. As projected by IEA (2021b), EVs sales are expected to experience a huge growth reaching about 70 million sales in 2040. Among

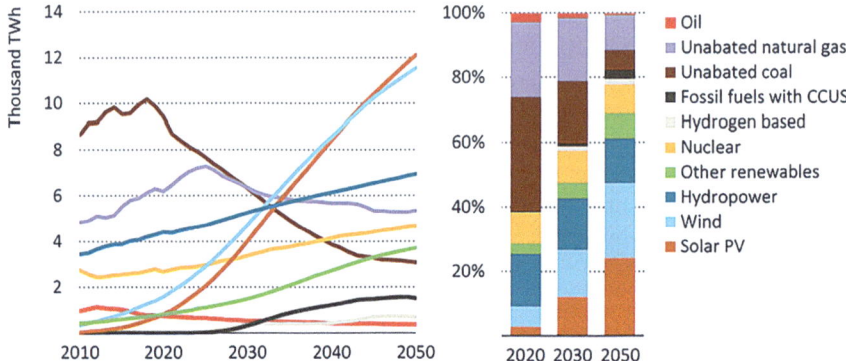

Fig. 3 Global electricity generation by source in the Announced Pledges Case (APC). *Note* Other renewables include geothermal, solar thermal and marine. Reprinted from IEA (2021a); license: CC BY 4.0

light-duty vehicles, the BEVs are expected to be predominant, accounting for around 80% of total sales by 2040 (IEA, 2021a).

Also, energy storage systems will have a pivotal role in the energy transition: batteries will allow the increase in flexibility and reliability of electricity grids, in which the share of energy generation from renewable and intermittent sources will be predominant. Globally, about 15.5 GW of energy storage capacity was provided as of 2020 by batteries connected to the electricity grid. Annual capacity additions are projected to reach 105 GW by 2040.

The expected global rise in electric cars and battery storage is shown in Fig. 4.

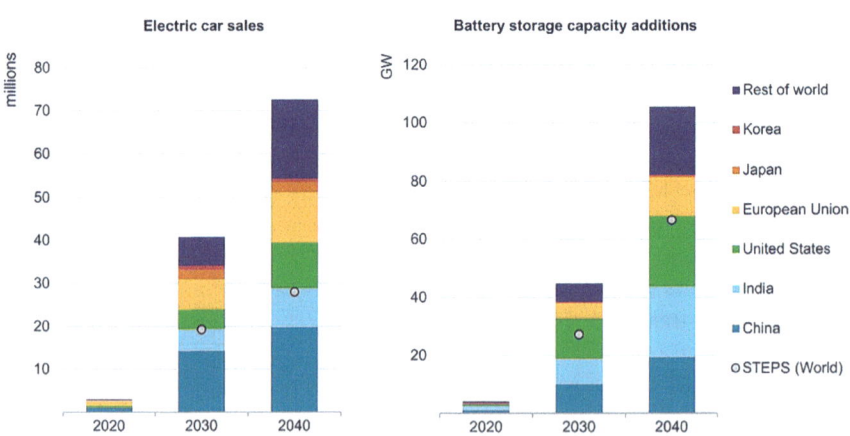

Fig. 4 Annual electric car sales and battery storage addition capacity in the IEA Sustainable Development Scenario [SDS]. *Note* Electric cars include battery electric and plug-in hybrid electric passenger light-duty vehicles. STEPS = Stated Policy Scenario. Reprinted from IEA (2021b); license: CC BY 4.0

Table 2 Critical and strategic raw materials used in the clean technologies selected for this study. Elaboration from IEA (2021a) and Carrara et al. (2023)

Material	Wind turbines	Photovoltaic modules	Electric motors	Li-ion Batteries
Aluminium	•	•	•	•
Antimony		•		
Arsenic		•		
Boron	•	•	•	
Cobalt				•
Copper	•	•	•	•
Fluorspar		•		•
Gallium		•		
Germanium		•		
Lithium				•
Manganese	•			•
Natural graphite				•
Nickel	•	•		•
Niobium	•			
Phosphorus		•		•
REE magnets	•		•	
Silicon metal	•	•	•	

Table 2 resumes the strategic and critical raw materials used in the four clean energy technologies identified in the scope of the present report, based on the studies of IEA (2021b) and Carrara et al. (2023). This further highlights the close connection between these technologies and the raw materials that could and should be recovered through recycling.

Competing Interests The authors have no conflicts of interest to declare that are relevant to the content of this chapter.

References

Binnemans, K., Jones, P. T., Blanpain, B., Van Gerven, T., Yang, Y., Walton, A., et al. (2013). Recycling of rare earths: A critical review. *Journal of Cleaner Production, 51*, 1–22. https://doi.org/10.1016/j.jclepro.2012.12.037.

Carrara, S., Alves Dias, P., Piazzotta, B., & Pavel, C. (2020). Raw materials demand for wind and solar PV technologies in the transition towards a decarbonised energy system. https://data.europa.eu, https://doi.org/10.2760/160859.

Carrara, S., Bobba, S., Blagoeva, D., Dias, A., Cavalli, P., Georgitzikis, A., et al. (2023). *Supply chain analysis and material demand forecast in strategic technologies and sectors in the EU-a foresight study.* JRC Science for Policy Report. EU CRM. https://doi.org/10.2760/334074.

European Commission. (2020). Study on the EU's list of Critical Raw Materials-final report (2020). https://op.europa.eu/en/publication-detail/-/publication/8dabb4c1-f894-11ea-991b-01aa75ed71a1.

European Commission. (2023a). COM(2023) 160 final. https://doi.org/10.2760/386650.

European Commission. (2023b). SWD(2023) 306 final. https://eur-lex.europa.eu/legal-content/EN/TXT/?uri=SWD%3A2023%3A306%3AFIN&qid=1684144015777.

European Commission, & EIP on Raw Materials. (2021). Raw materials scoreboard 2021. https://doi.org/10.2873/567799.

European Commission, & Joint Research Centre. (n.d.). RMIS–raw materials information system. https://rmis.jrc.ec.europa.eu/.

European Commission. (2015). COM(2015) 614 final. https://doi.org/10.1016/0022-4073(67)90036-2.

Gregoir, L., & van Acker, K. (2022). Metals for clean energy: Pathways to solving Europe's raw materials challenge. https://eurometaux.eu/metalscleanenergy.

Grohol, M., & Veeh, C. (2023). Study on the critical raw materials for the EU 2023–final report. https://op.europa.eu/en/publication-detail/-/publication/57318397-fdd4-11ed-a05c-01aa75ed71a1.

IEA. (2021a). Net zero by 2050. https://www.iea.org/reports/net-zero-by-2050.

IEA. (2021b). The role of critical minerals in clean energy transitions. https://www.iea.org/reports/the-role-of-critical-minerals-in-clean-energy-transitions.

IEA. (2023a). Energy technology perspectives 2023. https://www.iea.org/reports/energy-technology-perspectives-2023.

IEA. (2023b). Tracking clean energy progress 2023. https://www.iea.org/reports/tracking-clean-energy-progress-2023.

UNEP. (2011). Recycling rates of metals–a status report. https://www.unep.org/resources/report/recycling-rates-metals-status-report.

Lithium-Ion Batteries

Stefano Puricelli

Abstract This chapter explores the growth and technological advancements in the recycling of lithium-ion batteries (LIBs), a key component in the transition to sustainable energy solutions. Widely used in electric vehicles (EVs) and energy storage systems (ESSs), LIBs are valued for their high energy density, efficiency, and long lifecycle. The end-of-life management of these batteries is strategic, driven by the rising demand for raw materials and the need to mitigate environmental impacts. Recycling not only reduces dependency on mining but also addresses the potential future shortfall of critical resources like lithium, nickel, and manganese. The chapter updates the status of repurposing and recycling LIBs in Europe, also highlighting the evolving European Union regulations. By integrating advanced recycling methods and legislative frameworks, LIBs recycling plays a vital role in enabling a sustainable energy transition.

Keywords Li-ion battery · Electric vehicles · Energy storage system · Hydrometallurgy · Pyrometallurgy · Recycling

1 Introduction

An analysis conducted in 2022 by the McKinsey Battery Insights team forecast that the entire lithium-ion battery (LIB) chain, spanning from mining to recycling, is expected to experience a remarkable annual growth of over 30% from 2022 to 2030 (McKinsey & Company, 2023). This trajectory would culminate in a market size of 4.7 TWh by the end of the period. In fact, due to their low weight and high performance, the utilisation of LIBs plays not only a central role in small-scale applications (namely the consumer electronics), but also in large-capacity sectors, such as Electric Vehicles (EVs) and Energy Storage Systems (ESSs).

S. Puricelli (✉)
Department of Civil and Environmental Engineering, Politecnico di Milano, Milano, Italy
e-mail: stefano.puricelli@polimi.it

MatER Study Center, Laboratory for Energy and the Environment Piacenza, Piacenza, Italy

© The Author(s), under exclusive license to Springer Nature Switzerland AG 2025
M. Grosso and L. Rigamonti (eds.), *Waste Flows Generated by the Energy Transition*,
PoliMI SpringerBriefs, https://doi.org/10.1007/978-3-031-88951-6_2

The imperative for recycling is driven by the expected shortage of raw materials. If no further mining projects were developed, in 2030 the global expected difference between demand and supply would be around 1,730 kt for lithium (Li), 295 kt for nickel (Ni), and 1,590 for manganese (Mn) (McKinsey & Company, 2023). On the contrary, there could be around 45 kt of cobalt (Co) in excess, because of its expected future decreased presence in LIBs thanks to the diffusion of Lithium-Ferro-Phosphate (LFP) chemistry. Globally, from 2020 to 2040 there could be a 20% annual increase in battery materials for recycling (all battery types) (McKinsey & Company, 2023).

2 Characterisation of the Waste Flow

Lithium-based batteries have rapidly gained popularity and the field is still in active development. Compared to other battery technologies, LIBs offer one of the highest coulombic efficiencies (more than 99%), which is the ratio between charge and discharge capacity (Miao et al., 2019). Furthermore, they have a very high energy density, a long cycle life, fast and efficient charging with minimal energy wasted, a low self-discharge rate, no memory effect and require little maintenance. However, their safety is becoming a topic of concern after reports of the devices catching fire or explosion due to battery failure.

A LIB cell is composed by a shell, an electrolyte, a cathode, an anode, a separator and sealing parts. The cathode material defines the battery's specific characteristics, such as voltage and energy density. Alongside, the anode has the capacity to provide charged particles to the cathode during the discharge cycle. A permeable membrane known as the separator serves to avoid internal short circuiting, despite allowing the flow of Li ions between the cathode and the anode. This stream of particles is further facilitated by the electrolyte, a conductive substance that aids the movement of the ions. A LIB is a rechargeable battery in which Li ions move from the anode (negative electrode) to the cathode (positive electrode) during discharge, and back to the anode when charging. A LIB is generally composed of (Heelan et al., 2016; Lebedeva et al., 2016; Miao et al., 2019; Ordoñez et al., 2016; Winslow et al., 2018; Zeng et al., 2014; Zhang et al., 2018):

- casing: stainless steel, nickel-plated steel, aluminium (Al), plastic, Al-plastic
- cathode (positive electrode):
  - current collector: Al
  - active material: lithium cobalt oxide (LCO), lithium nickel manganese cobalt oxide (NMC), lithium nickel cobalt aluminium oxide (NCA), lithium manganese oxide (LMO), lithium iron phosphate (LFP)
  - binder: poly-vinylidene fluoride (PVDF), carboxymethyl cellulose, styrene butadiene rubber
  - additives.

- anode (negative electrode):
  - current collector: copper
  - active material: natural or synthetic graphite, lithium titanate (LTO), mesophase and amorphous carbon, silicon, alloys of lithium and metals (tin, silicon, aluminium), metallic nanoparticles, etc.
  - binder: poly-vinylidene fluoride (PVDF), carboxymethyl cellulose, styrene butadiene rubber, polytetrafluoroethylene
  - additives.
- electrolytic solution:
  - liquid:

    electrolyte: lithium salts (e.g., $LiPF_6$)
    solvent: propylene carbonate (PC), ethylene carbonate (EC), dimethyl carbonate (DMC), diethyl carbonate (DEC), dimethyl sulphoxide (DMSO), methyl ethyl carbonate (MEC).
  - solid:

    polymeric or ceramic.
  - additives
- separator: polyolefin materials, such as polyethylene (PE), polypropylene (PP), high density polyethylene (HDPE), ultrahigh molecular weight polyethylene (UHMWPE).

An example of LIB cell composition is shown in Table 1.

In terms of raw materials demand, Table 2 is based on the study of Carrara et al. (2023) and highlights the specific use of the identified material for the LIBs technology.

In a battery electric vehicle (BEV), a cell is the smallest component of a traction battery. Individual cells are connected into modules that, in turn, are jointed in a battery pack (Harper et al., 2019). Cells can have various shapes: cylindrical, prismatic or pouch.

A battery pack, in addition to the battery modules and the structural systems, comprises the battery management system (BMS), the thermal management system,

Table 1 Exemplifying LIB cell composition (Mossali et al., 2020)

Component	Mass (%)
Casing	20–26
Cathode	25–30
Anode	15–25
Electrolytic solution	10–15
Separator	4–10
Additives	Balance

Table 2 Selection of raw materials employed for Li-ion batteries (Carrara et al., 2023)

Use	Material
Cathode	• Aluminium as current collector foil, cathode material in NCA batteries, high purity alumina in coatings • Cobalt in LCO, NCA, and NMC batteries • Lithium • Manganese in NMC and LMO batteries • Nickel as hydroxide or intermetallic compounds in NMC and NCA batteries • Niobium in future cathode material (coatings) to improve stability and energy density
Anode	• Copper as current collector foil • Graphite • Lithium as Li metal in future anodes • Niobium in future anode material (coatings) to improve stability and energy density • Phosphorus in LFP batteries • Silicon in future anodes to enhance energy density • Titanium in future anode materials and coatings and in LTO batteries
Electrolytic solution	• Lithium as salt (electrolyte)
Rest of the battery	• Aluminium for battery casing • Copper in wires and other conductive parts • Titanium for battery casing

Note Critical raw materials are underlined

electrical systems, and a case. LIBs can be used both for EVs and stationary ESSs. For EVs, the volumetric energy density is a priority, while for ESSs the priority is given to the cost per kWh per cycle (IDTechEx, 2023). The gravimetric energy density of LIBs ranges from 90 to 260 Wh/kg, depending on the battery chemistry (IDB, 2024).

Research efforts are dedicated to new cell chemistries capable to outperform LIBs and address their critical issues:

- *Sodium-ion batteries* are developing as a feasible alternative substitution for LIBs, mainly because sodium is a low cost, abundant element and safer than lithium (Northvolt, 2024). As a drawback, sodium-ion batteries have lower volumetric energy density than LIBs. They could be used in low-range vehicles (e.g., two- or three-wheelers and city cars) or in ESSs (IDTechEx, 2023).
- *Lithium metal batteries* refer to batteries with Li as anode. They could offer a significant decrease in the overall battery mass, thanks to the ten-times higher theoretical specific capacity compared to a graphite anode (Lebedeva et al., 2016).
- *Lithium-sulphur batteries* are based on sulphur-containing cathodes and lithium anodes. The use of an abundant element like sulphur could lower the cost of batteries (Lebedeva et al., 2016).

- *Lithium-air batteries*, based on Li-metal and oxygen, they could offer a practical specific energy density of 500–1,000 Wh/kg, sufficient for 500 km of driving range in a BEV (Lebedeva et al., 2016).
- *Solid-state batteries* use solid electrolytes and could have higher energy density and quicker charging times (Lebedeva et al., 2016).
- *Vanadium redox flow batteries* have lower energy density and longer lifetime than LIBs (da Silva Lima et al., 2021). These batteries can be effectively utilised for ESSs.

3 Legislative Framework

In 2023, the European Union issued the Regulation (EU) 2023/1542 concerning batteries and waste batteries and introducing a regulatory framework for the battery life cycle as a whole (European Parliament & Council of the EU, 2023). The Regulation amends the Waste Directive 2008/98/EC and repeals the former Battery Directive 2006/66/EC. This Regulation introduces new provisions concerning the whole battery life cycle. Firstly, five battery categories are defined: portable batteries; EV batteries; industrial batteries (with stationary battery energy storage systems as a sub-category); light means of transport (LMT) batteries; starting, lighting and ignition (SLI) batteries. These battery categories are affected by specific requirements. The imposed targets, which will be gradually introduced within the next decade, concern the use of recycled materials in new batteries and the waste collection and recycling. Figure 1 shows the targets affecting the LIBs, which shall apply from 18 August 2025.

The previous Directive 2006/66/CE, which is repealed with effect from 18 August 2025, set a collection rate of 45% for all the batteries to be achieved by 2016. In 2022 the EU collection rate was 46% (Eurostat, 2023). In addition, the Directive established that recycling processes must achieve, by 2011, a minimum efficiency target of 65% for lead-acid batteries, 75% for nickel–cadmium batteries, and 50% for other batteries (the group of LIBs). The recycling efficiency means the "ratio obtained by dividing the mass of output fractions accounting for recycling by the mass of the waste batteries and accumulators input fraction expressed as a percentage" (Eurostat, 2023). The recycling efficiency target of 50% was reached by all EU countries which reported data in 2022 (Eurostat, 2023). The landfilling or incineration of industrial waste LIBs and waste LIBs for vehicles is forbidden.

The European List of Waste (ELW) code for waste LIBs is 16 06 05 ("other batteries and accumulators"). Surprisingly, they are not classified as hazardous waste, although LIBs contain flammable substances and residual charge, which can cause fire and explosions. According to the Agreement concerning the International Carriage of Dangerous Goods by Road, LIBs are considered as hazardous goods (IDB, 2024). Storage and transport of used and waste LIBs are controlled by several national and international regulations.

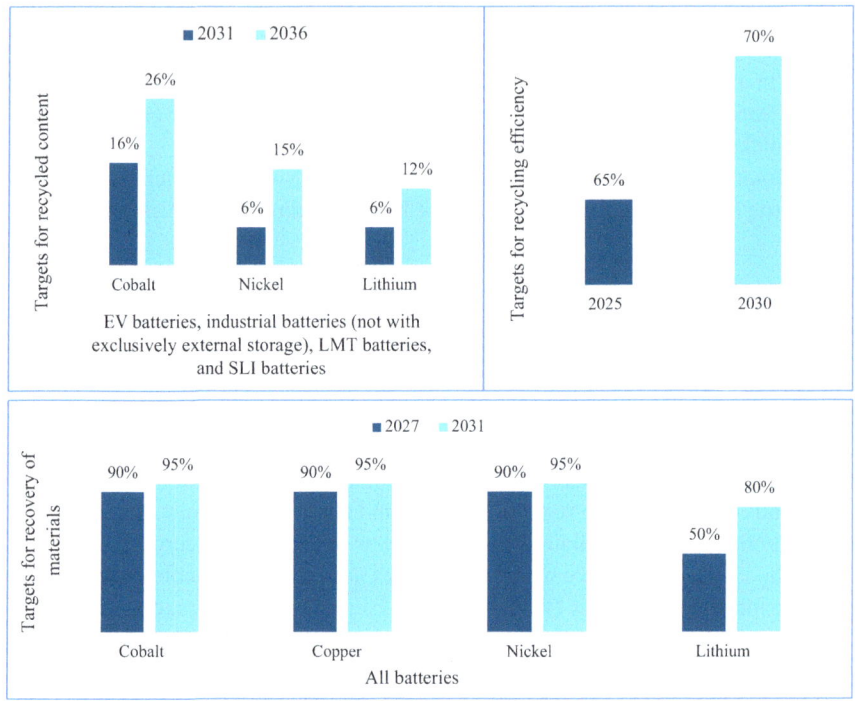

Fig. 1 Targets of regulation (EU) 2023/1542 for Li-ion batteries (European Parliament & Council of the EU, 2023)

4 End-of-Life Management Practices in a Circular Approach

Following a circular approach, three alternative scenarios can be implemented to exploit the full potential of used LIBs: reuse for the same purpose (reuse and remanufacturing), reuse for a different purpose (repurposing), and recycling. Before reusing or recycling a LIB, preliminary operations are necessary. Since many terms are commonly used, when describing these operations (e.g., remanufacturing, refurbishment, repairing, reconditioning) this study refers to the definitions from the Regulation (EU) 2023/1542 and the Directive 2008/98/EC:

- Directive 2008/98/EC (European Parliament & Council of the EU, 2008):
  - *'re-use' means any operation by which products or components that are not waste are used again for the same purpose for which they were conceived*;
  - *'preparing for re-use' means checking, cleaning or repairing recovery operations, by which products or components of products that have become waste are prepared so that they can be re-used without any other pre-processing*;

- 'recycling' means any recovery operation by which waste materials are reprocessed into products, materials or substances whether for the original or other purposes. It includes the reprocessing of organic material but does not include energy recovery and the reprocessing into materials that are to be used as fuels or for backfilling operations.

- Regulation (EU) 2023/1542 (European Parliament & Council of the EU, 2023):
  - 'repurposing' means any operation that results in a battery, that is not a waste battery, or parts thereof being used for a purpose or application other than that for which the battery was originally designed;
  - 'preparation for repurposing' means any operation, by which a waste battery, or parts thereof, is prepared so that it can be used for a different purpose or application than that for which it was originally designed;
  - 'remanufacturing' means any technical operation on a used battery that includes the disassembly and evaluation of all its battery cells and modules and the use of a certain number of battery cells and modules that are new, used or recovered from waste, or other battery components, to restore the battery capacity to at least 90% of the original rated capacity, and where the state of health of all individual battery cells does not differ more than 3% between cells, and results in the battery being used for the same purpose or application as the one for which the battery was originally designed;
  - 'preparation for recycling' means the treatment of waste batteries prior to any recycling process, including, inter alia, the storage, handling and dismantling of battery packs or the separation of fractions that are not part of the battery itself.
  - 'treatment' means any operation carried out on waste batteries after they have been handed over to a facility for sorting, preparation for re-use, preparation for repurposing, preparation for recycling or for recycling.

According to this set of definitions, the main EoL management options for LIBs are graphically shown in Fig. 2.

4.1 Repurposing

Repurposing means that the spent batteries are reconfigured for secondary utilisation in less demanding applications. Typically, the repurposing is very appropriate for LIBs from BEVs. When the battery's performance diminishes, no longer meeting the standards required for electric vehicles traction and range, it could still be qualified for less-stressful situation, such as energy storage systems (ESSs), before being recycled (Hua et al., 2021). Prolonging the service life has the potential to decrease the overall demand for raw materials, mitigating adverse impacts on the environment. The landscape of reuse applications encompasses mobile scenarios (e.g., two-wheelers, mobile charging robots) and stationary ones, such as renewable

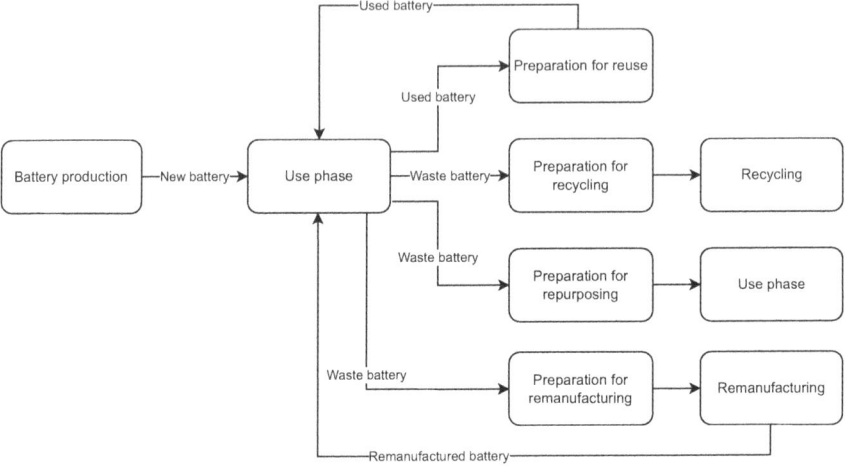

Fig. 2 Main EoL management practices for batteries

energy storage (e.g., wind and solar energy). In the latter case, repurposed batteries store excess generation and address the supply–demand imbalances through peak shaving and load shifting. The ESSs can also guarantee uninterrupted power in case of temporary disruptions in the power supply.

4.2 Remanufacturing

Remanufacturing is a well-known practice in the automotive sector, and it is the group of operations which return a used product to like-new condition (Ramoni & Zhang, 2013). The different stages of remanufacturing are typically the following ones: discharge, disassembly at the module/cell level, cleaning, inspection, health benchmark tests, substitution of damaged parts, reassembly, and testing (Bobba et al., 2018; European Parliament & Council of the EU, 2023; Ramoni & Zhang, 2013). While traditionally managed manually, efforts are underway to automate disassembly through artificial intelligence and machine vision sensors, ensuring safety, ergonomics, time and cost efficiency, and a controlled environment. Companies are nowadays operating disassembly manually or in semi-automated conditions where humans and robots collaborate (Harper et al., 2023). Remanufacturing faces a significant challenge with permanent joints used in battery manufacturing. Resistance welding, laser welding, or irreversible adhesive joints are some of the common methods used in battery manufacturing, but by contrast they cannot be completely disjointed without breakages. To achieve remanufacturing, a non-destructive disassembly is requested. For example, to remove the cover of a battery pack, the gasket must be destroyed and replaced with a new one (Kampker et al., 2024). Also, the removal of faulty battery modules requires the substitution of the layer of thermal gap filler, which is an adhesive material between the module and the cooling plate used to enhance the thermal conduction (ibid.). Recent research focuses on developing easily disassemblable joints to increase remanufacturing efficiency and to facilitate repairs. The changeability of the cells can be also enhanced thanks to a plug-in system that simplifies cells housing in the module and makes the battery periphery parts (e.g., sensors, cooling) easy to be connected.

4.3 Recycling

Recycling spent batteries allows not only to decrease the environmental impact and the CO_2 emissions associated with the extraction of raw materials, but it is somehow an unavoidable process due to various economic and social issues related to the extraction of virgin minerals. Furthermore, for LIBs, recycling is also economically attractive, due to their content of valuable metals. This paragraph provides a description of both established and experimental techniques to recover metals from EoL batteries, mainly focusing on LIBs.

Preparation for recycling does not alter the condition of the LIB cells. Typical operations are storage, dismantling of battery packs to module or cell level, sorting, separation of auxiliary components, and "non-invasive" discharging.

Owing to the complex assembly of LIB components and to the variety of materials used, a recycling strategy needs generally an optimised combination of mechanical, thermal, and hydrometallurgical treatments (not every step is required). Recycling processes can be classified in three routes (Fig. 3):

- mechanical/thermal treatments + hydrometallurgy;
- mechanical/thermal treatments + pyrometallurgy + hydrometallurgy;
- pyrometallurgy + hydrometallurgy.

Before applying any kind of recycling treatment, spent LIBs need to be deactivated and this can be done in different ways (Neumann et al., 2022). One technique consists in LIBs immersion in conductive solutions. A second way is to connect the batteries to an electrical circuit which can recover the residual energy content. Other ways for discharging employ thermal or cryogenic methods (Friedrich, 2017; Neumann et al., 2022). However, the cheapest and most widespread method is the in-process stabilisation during opening, which consists in shredding LIBs in an inert atmosphere (Harper et al., 2019).

Mechanical and physical treatments have the main purpose of recovering the plastics, the metallic components (e.g., cases, Al and Cu), and the so-called "*black mass*" (a mixture of cathode and anode active materials). Generally, after comminution, materials can be separated by a combination of physical separation processes e.g., sieves, filters, magnets, eddy current separators, and shaker tables (Harper et al., 2019). Graphite separation is also possible by flotation. To reduce the risks of fire

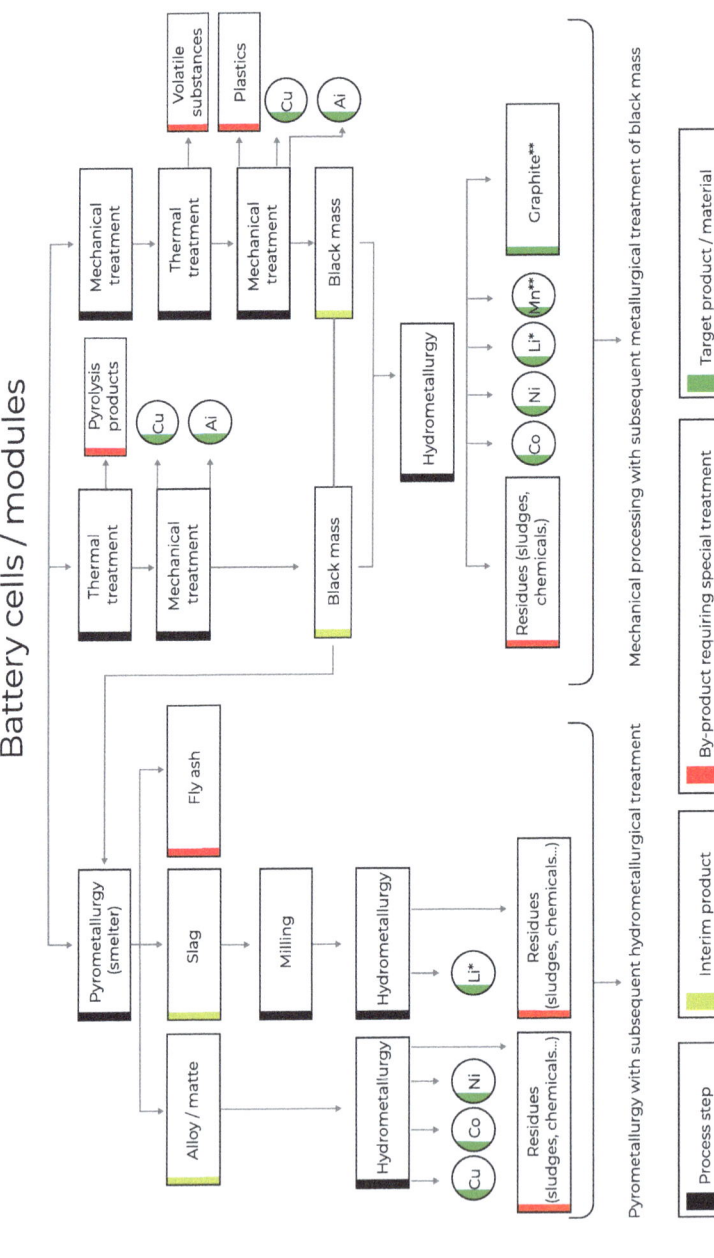

Fig. 3 Overview of LIBs recycling routes. Reprinted from IDB (2024)

and explosion, batteries must be fully discharged, or shredding can be performed in an inert atmosphere. Volatile substances (e.g., solvents) can be removed through heating and/or vacuum distillation (IDB, 2024).

The main targets of thermal treatments are the organic components contained in the spent batteries i.e., solvents and binder. The thermal process is carried out by heating the batteries to evaporate or decompose the solvents, the binder, and other volatile substances. The process can take place either in the presence of oxygen/air (incineration) or without oxygen (pyrolysis). With incineration all the organic components (plastics, solvent, binder) and graphite are destroyed. With pyrolysis it is possible to recover organic compounds from batteries by recondensation (Neumann et al., 2022). The oxygen-free environment requested for pyrolysis can be obtained by using nitrogen or by creating vacuum conditions (Harper et al., 2023).

Pyrometallurgical treatments can be conducted either with whole cells or modules, without the need for a special pre-treatment (Harper et al., 2019). In pyrometallurgy, a furnace at high temperature reduces the batteries to an alloy containing Co, Cu, and Ni. The byproduct of the process is the slag, containing, among others, Al and Li. Lithium can also be partially found in the fly ash. Fe mainly goes into the alloy, whilst Mn is distributed in both the alloy and the slag (Harper et al., 2023). Both the alloy and the slag can then be treated by hydrometallurgy for further recovery. The gases, instead, must be properly treated before their release in the atmosphere. Carbon and organic compounds are oxidised and not reclaimed.

Hydrometallurgy means the isolation and recovery of metals (from cathode and anode) at low temperatures by using aquatic media (Neumann et al., 2022). The first step is the leaching (acid leaching, bioleaching, etc.) of metallic components contained in the black mass. Graphite is the solid residue after filtration. Metal ions are then separated with a combination of techniques such as solvent extraction, ion exchange, chemical precipitation, and electrochemical deposition (ELIBAMA, 2014). Al, Fe, and Cu are firstly removed because considered as impurities. The target products are salts of Mn, Co, Ni, and Li.

A brief comment can be made on direct recycling, which is the retrieval and direct reuse of active materials in the same production chain (LIBs production) (Magni et al., 2024). In principle, it should be the preferred recycling method. Currently, due to the variety and contamination in LIBs, it is unrealistic for waste LIBs but could be feasible for scraps from battery production plants, which are a more uniform waste flow (IDB, 2024; Motus-E, 2023).

4.4 Performances, Benefits, Drawbacks

Recycling processes vary significantly, and many strategies have been tested at industrial scale. As already mentioned, hydrometallurgy is almost indispensable for isolating the precious metals contained in waste LIBs; also, pyrometallurgical processes ultimately employ hydrometallurgy for the treatment of the alloy and, possibly, the slag. Benefits and drawbacks of these two general metallurgical techniques are enlisted in Table 3.

LIBs recycling processes are mainly focused on the recovery of high value metals such as Co, Ni and Li (Zhao et al., 2021). In general, recycling costs depend on the type of LIB. In case of LFP or LIBs with low contents of Co and Ni, the value of recovered materials can be lower than the recycling costs (IDB, 2024). Nevertheless, the prices of these metals are affected by great volatility, making the estimate very uncertain and continuously variable. To achieve the affordability of recycling, therefore, the Extended Producer Responsibility (EPR) systems are a valid way to cover the costs.

Currently, the European recycling companies are using both pyrometallurgical and hydrometallurgical treatments, or a combination of the two (Fig. 3). Regarding the environmental impacts, the current consensus in literature is that pyrometallurgical recycling has higher impacts (Domingues & de Souza, 2024), because of high energy consumption, toxic emissions, and the impacts of the slag treatment (Harper et al., 2023). Hydrometallurgy reduces the overall impacts; since its burdens are related to the large demand for water and chemicals, the impacts can be reduced by reusing solvents and by-products (Neumann et al., 2022). Direct recycling has the lowest impacts and highest revenues, therefore is particularly suitable for LIB having low-value chemistries. However, a recent review showed divergent results in Life Cycle Assessments (LCAs), with the highest impacts for hydrometallurgy and a lack of comprehensive assessments for direct recycling (Domingues & de Souza, 2024). To

Table 3 Advantages and disadvantages of pyrometallurgy and hydrometallurgy (Friedrich, 2017; IDB, 2024; Motus-E, 2023)

Method	Advantages	Disadvantages
Pyrometallurgy	• Sorting and size reduction not necessary • LIBs can be recycled together with other wastes • Efficient recovery of Cu, Co, and Ni • Fe, Al, carbon, and organics used as reducing agent and/or energy carrier • High productivity, moderate area demand • No wastewater	• High temperature and energy demand • Recovery rates lower than in hydrometallurgy • Need for further hydrometallurgical refining • Recovery of Li only with dedicated plants for LIBs • Complex off-gas cleaning
Hydrometallurgy	• Low temperature and reasonable energy demand • Very high recovery rates • Recovery of carbon • Less off-gas	• Water requirement and complex wastewater treatment • Complexity of the process and high costs • Processes sensitive to the input black mass • Use of various chemical reagents • Sorting and size reduction necessary • Low kinetics, high area demand

enhance the precision of these assessments, future LCAs should include primary data for collection, pre-treatment and waste disposal, as well as cover a complete set of impact categories.

5 Industrial Initiatives

In this section the industrial initiatives for LIBs repurposing and recycling are synthetised.

5.1 Battery Repurposing

Currently operative Energy Storage Systems (ESSs) made with repurposed EV LIBs in Europe are at least 19 and are enlisted in Table 4. The reported capacity ranges from 0.1 to 13 MWh, while the number of connected individual battery packs ranges from 10 to 60, with a single case of 1,000. The reported number of connected battery modules from cars is between 1,800 and 2,600. Spent LIBs mostly come from retired electric passenger cars, but also from plug-in hybrid passenger cars and buses. These initiatives are frequently managed by the automakers themselves, in collaboration with specialised companies. An overview of the second life applications in Europe can be found in Battery-News (2023a).

5.2 Battery Recycling

Europe hosts at least 31 facilities dedicated to recycling lithium-ion batteries (LIBs), employing a variety of treatment methods. These plants are enlisted in Table 5, where only those actually operational are included. They have a wide range of capacities, from small-scale operations of 10 tonnes per year to large plants handling 25,000 tonnes per year. Fourteen plants only apply mechanical treatment by shredding, crushing, and sorting, to separate metals, plastics, and black mass. Seven plants are capable to treat the black mass through hydrometallurgy, for recovering Li-, Co-, Ni-, and Mn- based products. Four plants employ pyrometallurgy, always complemented by hydrometallurgy. In two cases a thermal pre-treatment is used to disintegrate the modules/cells or detach the coating of the electrode conductor foils and remove the separator and the electrolyte.

A few plants adopt integrated processes combining mechanical, thermal, and chemical treatments. These facilities focus on comprehensive material recovery, including both metals and graphite, achieving recycling efficiencies of up to 95%.

A broad overview of the European announced plants can be found in Battery-News (2023b).

Table 4 Selection of industrial initiatives repurposing spent EVs LIBs for energy storage purposes

Stakeholders	Location	Highlights
Audi, RWE	Herdecke, Germany	Storage facility consisting of 60 used LIBs from Audi e-tron development vehicles and with a capacity of 4.5 MWh (RWE, 2021)
Audi, The Mobility House	Berlin, Germany	A 1.6-MWh storage system uses LIBs from used test vehicles to test interaction scenarios between EVs and the energy grid (The Mobility House, 2024)
Bosch, BMW, Vattenfall	Hamburg, Germany	2,600 Battery modules from 100 BMW EVs (ActiveE and i3 models) were tested and reassembled in a stationary storage system with a power output of 2 MW and an installed capacity of 2.8 MWh (Bobba et al., 2018)
Comsys, Fortum, Volvo Cars	Landafors, Sweden	A battery energy storage facility incorporating 48 PHEV Volvo batteries (Comsys, 2021). Capacity of 250 kWh and power of 1 MW
Connected Energy	–	Energy storage system (E-STOR systems) made with 24 Renault batteries, with a nominal power of up to 300 kW and a usable capacity of up to 325 kWh (Connected Energy, 2024)
Daimler, Mercedes-Benz Energy, Enercity	Hanover, Germany	Stationary accumulator made with 1,800 modules, with an output of 5 MW (Mercedes-Benz, 2017). The system is designed to be a spare parts store which held used LIBs under charge and, at the same time, balance the power market. The completion of the accumulator, reaching 3,240 battery modules, will increase the capacity to 17.4 MWh
Daimler, The Mobility House, GETEC, REMONDIS	Lünen, Germany	1,000 LIBS from smart fortwo electric drive cars were repurposed and used in a 13-MWh unit that stores renewable electrical energy (Mercedes-Benz, 2016)
Daimler, Mercedes-Benz Energy, The Mobility House, GETEC ENERGIE	Elverlingsen, Germany	A 9.8-MWh storage plant, made with 1,920 modules and with a power output of 8.96 MW (Mercedes-Benz, 2018). As the plant in Hannover, the plant is designed as a living replacement parts store
Eaton, Nissan	–	Energy storage systems that combine new and used Nissan LEAF batteries. Storage capacity for domestic applications: 4.2–10.1 kWh (Eaton, 2024b). Storage capacity for commercial/industrial applications: 20 kWh-5 MWh (Eaton, 2024a). The largest application is a 3-MW ESS installed in the Johan Cruijff Arena (Amsterdam) for back-up power, peak shaving, and storage of solar energy (Bobba et al., 2018; Motus-E, 2023)

(continued)

Table 4 (continued)

Stakeholders	Location	Highlights
EcarACCU	Zwaag, Netherlands	Used LIBs of EVs are disassembled into cell/module level and evaluated (EcarACCU, 2024). The approved battery cells are sold to specialised partners. The leftover materials are distributed to recycling centres or reused
Enel, Endesa, Nissan, Loccioni	Melilla, Spain	Energy storage system made of 78 Nissan LIBs, of which 30 new and 48 used (Enel Spa, 2022). Power of 4 MW and capacity of 1.7 MWh. LIBs are put directly in the system without disassembling and their lifetime is extended by six years (Enel X S.r.l., 2024)
GAIA	Flins, France	In its Battery Repair Expert Center (CERBF), GAIA repairs the batteries or prepares them for other uses, such as stationary energy storage (GAIA, 2024)
Jaguar Land Rover, Wykes Engineering	UK	A battery energy storage system utilising 30 I-PACE batteries and storing 2.5 MWh (Jaguar Land Rover, 2023). Batteries from prototype and engineering test vehicles
Jaguar Land Rover, Pramac	–	A portable energy storage unit made of I-PACE batteries from prototype and engineering test vehicles, with a capacity of up to 125 kWh (Jaguar Land Rover, 2022)
MAN Truck & Bus, VHH, Volkswagen Group	Hamburg, Germany	Energy storage system containing 50 batteries from VW Passat GTE with a total capacity of 495 kWh (MAN, 2019). The system reduces the peak load when charging electric buses
Mercedes-Benz Energy, Daimler Buses	Hanover, Germany	Stationary energy storage unit with a 500-kWh capacity, consisting of 28 used LIBs from Mercedes-Benz eCitaro buses (Mercedes-Benz Buses, 2023)
Mobilize, Morbihan Energies	Belle-Île, France	Stationary energy storage unit of 200 kWh capacity, incorporating 10 Renault Zoe second-life batteries (Mobilize, 2022)
Renault, The Mobility House	Douai, France	Battery storage made of first- and second-life batteries from Renault Zoe, with an output power of 4.7 MW (Renault Group, 2021)
Tauron Polska Energia	Jaworzno, Poland	Stationary energy storage system made of four battery modules from a Solaris electric bus, with a total capacity of 150 kWh (Electrive, 2023)

Table 5 European active industrial initiatives for recycling of spent LIBs in Europe

Stakeholders	Location	Process	t LIBs/y	Highlights	Sources
ACCUREC Recycling	Germany	T + M + H	3,800	Automotive/industrial Li-ion battery packs EV are dis-connected to <60 V and the cooling systems are drained. The batteries packs are dismantled to module/cell level. Al, Cu, plastics and steel are recovered. Household LIBs packs and cells are manually sorted by chemistry. The modules/cells are thermally disintegrated. Further comminution and multistep mechanical treatment remove steel, Cu, Al, and black mass. Black mass undergoes a hydrometallurgical process that recovers lithium carbonate and Ni-Co concentrate	ACCUREC Recycling (n.d., 2020)
AkkuSer	Finland	M	4,000	LIBs undergo a two-stage crushing, a magnetic separation and a mechanical separation. No preventive dis-charge is needed. The outputs are plastics, cardboard, and metal concentrates that are sent to external refining	Akkuser (n.d.), Lv et al. (2018)

(continued)

Table 5 (continued)

Stakeholders	Location	Process	t LIBs/y	Highlights	Sources
Duesenfeld, TU Bergakademie Freiberg	Germany	M + H	2,900 (Duesenfeld)	LIBs are discharged and the electrical energy is self-consumed. The disassembly recovers housing, BMS, screws, cables, and colling system. Batteries are then comminuted under an inert atmosphere. The solvent is recovered by a low-temperature vacuum dryer. Shredded materials are separated into plastics, Fe, Cu, Al, and black mass. The hydrometallurgical process is made in a pilot plant. Black mass is leached, and metals are separated from graphite. Various extraction methods are applied to recover Li, Co, Ni, and Mn salts. The overall recycling rate is 91%, with the black mass processing	ACCUREC Recycling (2020), Duesenfeld (n.d.), VDI Zentrum Ressourceneffizienz (2023)
Ecobat	UK Germany	M	500 (UK) 3,000 (Germany)	Battery packs are collected, firstly evaluated through visual checks and safety tests, then dis-mantled. Modules, subassemblies and cells are crushed and sorted to recover Cu, Al, steel, plastics, and the black mass	Battery-News (2023b), Ecobat (Ecobat, n.d.)
ERM	UK	M	2,000	LIBs are tested to assess if they can be repaired, reused, or recycled. LIBs are also discharged and dismantled	Electrive (2024)
ERM, Northvolt	Germany	M	10,000	Battery packs are discharged and dismantled. The high-grade copper and aluminium are recovered	ERM (2023)

(continued)

Table 5 (continued)

Stakeholders	Location	Process	t LIBs/y	Highlights	Sources
Fortum	Ikaalinen (FI) Kirchardt (DE) Harjavalta (FI)	M (Ikaalinen, Kirchardt) + H (Harjavalta)	3,000 (Ikaalinen) 3,000 (Kirchardt)	LIBs are disassembled and mechanically treated. Plastics, Al, Cu, and black mass are recovered. The black mass undergoes the hydrometallurgical processing and Ni-, Co-, Mn-, and Li-based products are recovered. The maximum recycling efficiency is 80%	Battery-News (2023b), Fortum (n.d.)
Hydrovolt	Norway	M	12,000	Battery packs are discharged, and the collected electrical energy powers the plant and supports the local grid. Battery packs are dismantled, modules are crushed and shredded. The plant enables up to 95% of metals from LIBs to be recovered, and it is powered 100% by renewable energy. The plant produces black mass, which is then sent to other companies	Hydrovolt (n.d.)
KYBURZ	Switzerland	M	200	A mechanical direct recycling process allowing to recover as much as 91% of the battery materials	Battery-News (2023b), KYBURZ (n.d., 2020)
Li-Cycle	Germany	M	10,000	Batteries are submerged and shredded. The process allows processing full EV and energy storage battery packs, without the need for disassembly, discharging, or thermal processes. Furthermore, it can recycle all types of lithium-ion battery materials. The process generates black mass and mixed Cu/Al	Li-Cycle (n.d., 2024)

(continued)

Table 5 (continued)

Stakeholders	Location	Process	t LIBs/y	Highlights	Sources
Mercedes-Benz	Germany	M + H	2,500	Integrated mechanical-hydrometallurgical process with an expected 96% recovery rate and supplied with 100% renewable electricity. The mechanical stage recovers plastics, Cu, Al, and Fe. The hydrometallurgical stage operates at up to 80 °C and extracts Co, Ni, and Li from the black mass	Mercedes-Benz (2024)
Nickelhütte Aue	Germany	P + H	4,000	LIBs are disassembled and transported toward a drum furnace where LIBs are melted, and an alloy is obtained. The alloy is comminuted and processed through pressure oxidation leaching, followed by solvent extraction. The plant can produce, for example, nickel sulphate, cobalt sulphate, and copper sulphate	ACCUREC Recycling (2020), Neumann et al. (2022), Nickelhütte Aue (2023)

(continued)

Table 5 (continued)

Stakeholders	Location	Process	t LIBs/y	Highlights	Sources
Primobius	Germany	M + H	10 (t/day)	The plant utilises a two-stage shredding process, followed by sorting, drying, and beneficiation. Plastics, steel casings, and foils of Co and Al are sold as scrap to existing metal and plastic recyclers. The black mass is processed in the hydro-metallurgical section consisting of leaching, purification, and precipitation. The refined products, including Co sulphate, Ni sulphate, Cu sulphate, along with Mn sulphate solution and Li sulphate solution, are suitable for sale back into the battery supply chain or for other industrial uses, while the remaining graphite anode material is dried and sold separately. Liquid ammonium sulphate solution is concentrated, crystallised, and sold as fertiliser	Primobius (n.d.-a, n.d.-b)
POSCO, SungEel HiTech	Poland	M	7,000 (produced black)	The battery recycling plant collects and disassembles scrap and used batteries from European battery manufacturers, producing black mass, which is sent to Korea for extraction of cathode materials such as Li, Ni, Mn, and Co	POSCO (n.d., 2022)

(continued)

Table 5 (continued)

Stakeholders	Location	Process	t LIBs/y	Highlights	Sources
Redux Recycling	Germany	T + M	10,000	From the discharge, the electric energy is recovered. Manual dismantling separates plastics, cables, Al, electronic components. Cells are pretreated through a thermal process, the coating of the electrode conductor foils is detached, and the separator and electrolyte are removed. The mechanical process isolates the active materials, a ferromagnetic fraction, an Al fraction, and an Al-Cu fraction. The recovered active materials are sent to external pyrometallurgical and/or hydro-metallurgical processes	REDUx Recycling (2018, 2024)
SNAM	Saint Quentin Fallavier (FR), Viviez (FR)	M + P + H	300 (Saint Quentin Fallavier)	Battery packs from automotive and aeronautics industries are dismantled, by separating cells from cables, electronic boards, boxes, protections, and other by-products. The batteries unsuitable for a second life are recycled. Portable batteries are sorted by type. The mechanical unit separates metallic fractions and plastics and paper. Then, materials undergo thermolysis in a low-O_2 atmosphere, where electrolytes and organic residues are removed. The distillation phase separates the various metals which are refined through a thermal treatment. The recovered metals are purified by hydrometallurgy. The recycling efficiency of LIBs is higher than 73%. The products of recycling are Al, Cd, Co, Cu, steel, Ni, electronic components, and plastics	Automobile Propre (2018), Lebedeva et al. (2016), SNAM (n.d., 2023)

(continued)

Table 5 (continued)

Stakeholders	Location	Process	t LIBs/y	Highlights	Sources
Stena Recycling	Sweden	M	10,00	LIBS, after testing, are prepared for second life or recycling. In the recycling process, LIBS are shredded in O_2-free environment. The electrolyte is separated by drying. A mechanical process sorts out plastics, Al, Fe and Cu. Black mass is delivered to industrial partners	Stena Recycling (n.d.)
SungEel HiTech	Hungary	M + H	7,000 (Szigetszentmiklós) 25,000 (Bátonyterenye)	LIBs are discharged, dismantled, thermally treated, and crushed. The resulting powder is processed with hydrometallurgy to recover cobalt sulphate, nickel sulphate, lithium carbonate, manganese sulphate, electrolytic nickel and electrolytic copper. It is not clear if the black mass is treated in Hungary or abroad (like for the POSCO plant)	Battery-News (2023b), Hungarian Investment Promotion Agency (2021), SungEel HiTech (n.d.)
TES	France	M + H	2,200	LIBs are shredded, grinded, and sieved to recover plastics, Cu, Al, and steel. Black mass is processed by hydrometallurgy to recover Co and Li products, but updated information is not available	ACCUREC Recycling (2020), Battery-News (2023b), Lv et al. (2018), TES (n.d.)

(continued)

Table 5 (continued)

Stakeholders	Location	Process	t LIBs/y	Highlights	Sources
Umicore	Belgium, Germany	P + H	7,000 (BE)	After dismantling of packs until module or cell level, no crushing or shredding is needed. After a treatment in a shaft furnace, a metal alloy (containing Co, Ni, and Cu) and a Li concentrate are generated. Plastics, electrolyte, and graphite are burnt off. The alloy and the Li concentrate are upgraded in hydrometallurgical processes for recovering metals in battery-grade quality. The recovery yields are over 95% for Ni, Cu, and Co and over 79% for Li	ACCUREC Recycling (2020), Umicore (2022, 2023)
Veolia-SARPI	France	M (Euro Dieuze Industrie) + H (CEDILOR)	1,000 (EDI) 400 (CEDILOR)	The recycling of EV LIBs consists of: (i) diagnosis and dis-charging; (ii) manual dismantling of plastic or Al casing, electronic components, wires, connectors, and the cooling system; (iii) cutting of modules, removing of Al protection and uncovering of cells; (iv) grounding up of cells and separation of black mass, paper and plastics, as well as Al, Cu and steel; (v) hydrometallurgical treatment of the black mass through mixing with reagents, separation, purification and concentration of metallic salts (Li, Ni, Co), which are sold to a metallurgist for refining and production of precursors	ELIBAMA (2013), Motus-E (2023), Muller et al. (2021), VEoLia (2024)

(continued)

Table 5 (continued)

Stakeholders	Location	Process	t LIBs/y	Highlights	Sources
Veolia-SARPI	Switzerland	P	500	Detailed information is reported after the table	Batrec Industrie (n.d.), Battery-News (2023b), Motus-E (2023)
Volkswagen	Germany	M	1,500	Battery packs are tested and discharged. Housings and attachments are removed. Usable modules are saved for second-life applications. Batteries are shredded, granules are dried and liquid electrolyte pumped out. The dried material is sifted, and the black mass separated. Magnetic and non-magnetic metals are separated. The resulting bags contain plastic bits, mixed Al and Cu, and the black mass	Volkswagen (2021)

Note M = mechanical, H = hydrometallurgical, P = pyrometallurgical, T = thermal

Fig. 4 Overview of the mechanical process for Li-ion batteries (photo courtesy of Batrec Industrie AG)

Case Study: Batrec Industrie. Here a focus on the state-of-the-art of LIBs recycling in Europe is made based on the Batrec Industrie AG, a Swiss company located in Wimmis, specialised in the pyrometallurgical treatment of primary batteries for the recovery of ferromanganese and zinc. Since November 2023, a new process for LIBs has been introduced (Figs. 4 and 5).

Batrec is processing 800 t of LIBs from the Swiss Market in 2024, mostly from portable batteries, with only a small portion coming from electric vehicles. The recycling process of LIBs is based on a mechanical treatment which starts with the pre-shredding of modules and cells down to 24 mm and then 14 mm, while the big battery packs are sent to other companies. The discharging of LIBs is achieved through the immersion into water, which prevents their ignition by cooling and dissipates the residual electricity. Water is used also to prevent risks during the shredding and the storage of shredded LIBs, which are then subject to an air-drying step. The shredded LIBs are then loaded to a hammer mill to enhance the liberation of black mass. The milled batteries undergo a series of mechanical steps through different machines: sieve, magnetic separator (to separate the steel), zig zag sifter (to separate heavy from light materials), balling mill (to achieve the granulation of copper and aluminium and enhance further separation). After that, three sieves separate the black mass from a mixture consisting of copper, aluminium, and light plastics. Three different separating tables separate the copper from the mixture of aluminium and plastics. The outputs of the process are steel, copper, aluminium mixed with light plastics, black mass of first quality and second quality, separating plastic foils, and

Fig. 5 Selection of photos regarding the mechanical recycling of LIBs (photo courtesy of Batrec Industrie AG)

heavy plastics. The copper and the mixture of aluminium and plastics are obtained in three size ranges: 1–3 mm, 0.5–1 mm, and 0.25–0.5 mm, while the black mass is the finest fraction. Black mass represents about 50% of the output and contains Li from the cathode, while Li from the electrolytic solution, in principle, could be recovered from wastewater produced during the pre-shredding. All the output materials are sent to external recycling, preparation for recycling, or disposal.

6 Conclusions

Li-ion batteries cells are composed of casing, cathode, anode, electrolytic solution, and separator. Individual cells can be connected into modules and packs, together with electrical and electronic systems, thermal management systems, and an external case. LIBs contain several critical and strategic raw materials in the cathode and anode, for example aluminium, cobalt, copper, graphite, lithium, manganese, nickel, and phosphorus. In Europe, LIBs are controlled by the Regulation (EU) 2023/1542 that sets targets for: (i) recycling efficiency of used LIBs; (ii) recovery of cobalt, copper, nickel, and lithium from used LIBs; (iii) recycled content of cobalt, nickel, and lithium in new LIBs. The European List of Waste code is 16 06 05, surprisingly not hazardous, although LIBs can cause fire and explosions and are already controlled, in this sense, by several regulations about storage and transport. The remanufacturing of used LIBs is under research, but still not a reality. Used automotive LIBs can be reconfigured for secondary utilisations and at least 19 stationary storage plants are operative in Europe, especially in Germany. European recycling companies are employing both pyrometallurgical and hydrometallurgical treatments, sometimes coupled. The most common approach, however, is a series of mechanical treatments that eventually separate the so-called black mass (a mixture of cathode and anode materials), to be sent to specialised plants. This review found 31 operative recycling plants in Europe, especially in Germany.

Competing Interests The authors have no conflicts of interest to declare that are relevant to the content of this chapter.

References

ACCUREC Recycling. (n.d.). ACCUREC recycling. https://accurec.de.
ACCUREC Recycling. (2020). Comparative study of Li-ion battery recycling processes. https://accurec.de/wp-content/uploads/2021/04/Accurec-Comparative-study.pdf
Akkuser. (n.d.). Recycling of high-grade cobalt Li-ion batteries. https://www.akkuser.fi/en/process-descriptions/high-grade-cobalt-li-ion-battery/.
Automobile Propre. (2018). Recyclage des batteries: Notre visite au cœur d'une usine française. https://www.automobile-propre.com/recyclage-des-batteries-notre-visite-au-coeur-dune-usine-francaise/.
Batrec Industrie. (n.d.). Battery recycling. Retrieved March 28, 2024, from https://batrec.ch/battery-recycling/portable-and-consumer-batteries/.
Battery-News. (2023a). Europe 2nd life application. https://battery-news.de/en/europe-2nd-life-application/.
Battery-News. (2023b). Europe battery recycling. https://battery-news.de/en/europe-battery-recycling/.
Bobba, S., Podias, A., Di Persio, F., Messagie, M., Tecchio, P., Cusenza, M. A., et al. (2018). *Sustainability assessment of second life application of automotive batteries.* JRC Technical Reports. https://doi.org/10.2760/53624.
Carrara, S., Bobba, S., Blagoeva, D., Dias, A., Cavalli, P., Georgitzikis, A., et al. (2023). *Supply chain analysis and material demand forecast in strategic technologies and sectors in the EU-a foresight study.* JRC Science for Policy Report. EU CRM. https://doi.org/10.2760/334074.
Comsys. (2021). Comsys delivers turnkey battery energy storage facility to Fortum. https://comsys.se/news/comsys-delivers-turnkey-battery-energy-storage-facility-to-fortum/.
Connected Energy. (2024). E-STOR battery energy storage. https://connected-energy.co.uk/battery-energy-storage/.
da Silva Lima, L., Quartier, M., Buchmayr, A., Sanjuan-Delmás, D., Laget, H., Corbisier, D., et al. (2021). Life cycle assessment of lithium-ion batteries and vanadium redox flow batteries-based renewable energy storage systems. *Sustainable Energy Technologies and Assessments, 46*, 101286. https://doi.org/10.1016/j.seta.2021.101286
Domingues, AM., de Souza, R. G. (2024). Review of life cycle assessment on lithium-ion batteries (LIBs) recycling. *Next Sustainability 3*, 100032. https://doi.org/10.1016/j.nxsust.2024.100032
Duesenfeld. (n.d.). Ecofriendly recycling of lithium-ion batteries. Retrieved April 10, 2024, from https://www.duesenfeld.com/recycling_en.html.
Eaton. (2024a). Energy storage systems. https://www.eaton.com/gb/en-gb/products/energy-storage.html.
Eaton. (2024b). xStorage home. https://www.eaton.com/it/it-it/markets/residential/il-sistema-all-in-one-di-eaton.html.
EcarACCU. (2024). Second-life EV battery cells. https://ecaraccu.nl/second-life/.

Ecobat. (n.d.). Lithium services. Retrieved March 29, 2024, from https://ecobat.com/our-business/ecobat-solutions/lithium-services/.

Electrive. (2023). Tauron builds second-life energy storage unit in Poland. https://www.electrive.com/2023/12/21/tauron-builds-second-life-energy-storage-unit-in-poland/.

Electrive. (2024). EMR opens EV battery recycling facility in Birmingham. https://www.electrive.com/2024/09/26/emr-opens-ev-battery-recycling-facility-in-birmingham/.

ELIBAMA. (2013). Euro Dieuze Industrie. https://elibama.wordpress.com/2013/01/03/173427237/.

ELIBAMA. (2014). Li-ion batteries recycling. https://elibama.wordpress.com/wp-content/uploads/2014/10/v-d-batteries-recycling1.pdf

Enel X S.r.l. (2024). Cos'è una batteria di seconda vita: Significato e procedimento. https://corporate.enelx.com/it/question-and-answers/what-is-second-life-battery.

Enel Spa. (2022). Enel lancia l'innovativo sistema di stoccaggio "Second Life" per batterie usate delle auto elettriche a Melilla in Spagna. https://www.enel.com/it/media/esplora/ricerca-comunicati-stampa/press/2022/03/enel-lancia-linnovativo-sistema-di-stoccaggio-second-life-per-batterie-usate-delle-auto-elettriche-a-melilla-in-spagna-.

ERM. (2023). EMR and Northvolt establish electric vehicle battery recycling facility in northern Germany. https://uk.emrgroup.com/find-out-more/latest-news/Northvolt-and-EMR-establish-electric-vehicle-battery-recycling-facility-in-northern-Germany.

European Parliament and Council of the EU. (2023). Regulation (EU) 2023/1542 of the European Parliament and of the Council concerning batteries and waste batteries. *Official Journal of the European Union, L 191.* https://eur-lex.europa.eu/eli/reg/2023/1542/oj.

European Parliament and Council of the EU. (2008). Directive 2008/98/EC. *Official Journal of the European Union, L 312.* https://doi.org/10.5040/9781782258674.0028.

Eurostat. (2023). Waste statistics-recycling of batteries and accumulators. https://ec.europa.eu/eurostat/statistics-explained/index.php?title=Waste_statistics_-_recycling_of_batteries_and_accumulators.

Fortum. (n.d.). Lithium-ion battery recycling technology. Retrieved April 2, 2024, from https://www.fortum.com/services/battery-recycling/lithium-ion-battery-recycling-technology.

Friedrich, B. (2017). State of research on Li-ion battery recycling. Kraftwerk Batterie, April, 25. https://doi.org/10.13140/RG.2.2.30525.26085.

GAIA. (2024). GAIA. https://www.gaiaautorecycling.com/en/gaia-pioneer-of-the-automotive-circular-economy/.

Harper, G., Sommerville, R., Kendrick, E., Driscoll, L., Slater, P., Stolkin, R., et al. (2019). Recycling lithium-ion batteries from electric vehicles. *Nature, 575*(7781), 75–86. https://doi.org/10.1038/s41586-019-1682-5

Harper, G. D. J., Kendrick, E., Anderson, P. A., Mrozik, W., Christensen, P., Lambert, S., et al. (2023). Roadmap for a sustainable circular economy in Lithium-ion and future battery technologies. *Journal of Physics: Energy, 5*(2), 021501. https://doi.org/10.1088/2515-7655/acaa57

Heelan, J., Gratz, E., Zheng, Z., Wang, Q., Chen, M., Apelian, D., et al. (2016). Current and prospective Li-ion battery recycling and recovery processes. *JOM Journal of the Minerals Metals and Materials Society, 68*(10), 2632–2638. https://doi.org/10.1007/s11837-016-1994-y

Hua, Y., Liu, X., Zhou, S., Huang, Y., Ling, H., & Yang, S. (2021). Toward sustainable reuse of retired Lithium-ion batteries from electric vehicles. *Resources, Conservation and Recycling, 168,* 105249. https://doi.org/10.1016/j.resconrec.2020.105249

Hungarian Investment Promotion Agency. (2021). SungEel Hitech is opening a new global green battery recycling plant in Bátonyterenye. https://hipa.hu/news/sungeel-hitech-hungary-is-opening-a-new-global-green-battery-recycling-plant-in-batonyterenye/.

Hydrovolt. (n.d.). Hydrovolt. Retrieved April 2, 2024, from https://www.hydrovolt.com/en.

IDB. (2024). Recycling and reuse of lithium batteries in Latin America and the Caribbean-analytical review of global and regional practices. https://publications.iadb.org/en/recycling-and-reuse-lithium-batteries-latin-america-and-caribbean-analytical-review-global-and.

IDTechEx. (2023). Sodium-ion batteries 2024–2034: Technology, players, markets, and forecasts. https://www.idtechex.com/en/research-report/sodium-ion-batteries-2024-2034-technology-players-markets-and-forecasts/978.

Jaguar Land Rover. (2022). Jaguar Land Rover gives second life to I-PACE batteries. https://media.jaguar.com/news/2022/03/second-life-jaguar-i-pace-batteries-power-zero-emission-energy-storage-unit.

Jaguar Land Rover. (2023). JLR creates new renewable energy storage system from used car batteries. https://media.jaguarlandrover.com/news/2023/08/jlr-creates-new-renewable-energy-storage-system-used-car-batteries.

Kampker, A., Heimes, H. H., Frieges, M., Klohs, D., Mussehl, V., Gross, L., et al. (2024). EV life cycle optimization through battery repair. https://www.pem.rwth-aachen.de/go/id/hwpp/lidx/1.

KYBURZ. (2020). KYBURZ battery recycling anlage. https://www.youtube.com/watch?v=HL6xscxzEJ8.

KYBURZ. (n.d.). Battery recycling. Retrieved April 22, 2024, from https://kyburz-switzerland.ch/en/battery-recycling.

Lebedeva, N., Di Persio, F., & Boon-Brett, L. (2016). *Lithium ion battery value chain and related opportunities for Europe*. https://doi.org/10.2760/6060.

Li-Cycle. (n.d.). Technology. Retrieved March 30, 2024, from https://li-cycle.com/technology/.

Li-Cycle. (2024). Li-cycle receives approval for government grant from State of Saxony-Anhalt for Lithium-ion battery recycling facility. https://li-cycle.com/press-releases/li-cycle-receives-approval-for-government-grant-from-state-of-saxony-anhalt-for-lithium-ion-battery-recycling-facility/.

Lv, W., Wang, Z., Cao, H., Sun, Y., Zhang, Y., & Sun, Z. (2018). A critical review and analysis on the recycling of spent Lithium-ion batteries. *ACS Sustainable Chemistry and Engineering, 6*(2), 1504–1521. https://doi.org/10.1021/acssuschemeng.7b03811

Magni, M., Colledani, M., & Harper, G. (2024). Editorial: The challenge towards more sustainable lithium ion batteries: From their recycling, recovery and reuse to the opportunities offered by novel materials and cell design. *Frontiers in Chemistry, 12*, 1421434. https://doi.org/10.3389/FCHEM.2024.1421434

MAN. (2019). New lease of life for vehicle batteries: pilot project started by MAN, Verkehrsbetriebe Hamburg-Holstein and Volkswagen. https://press.mantruckandbus.com/corporate/new-lease-of-life-for-vehicle-batteries-pilot-project-started-by-man-verkehrsbetriebe-hamburg-holstein-and-volkswagen/.

McKinsey & Company. (2023). Lithium-ion battery demand forecast for 2030. https://www.mckinsey.com/industries/automotive-and-assembly/our-insights/battery-2030-resilient-sustainable-and-circular.

Mercedes-Benz Buses. (2023). Kick-off in Hanover: Stationary energy storage system consisting of used eCitaro batteries goes into operation. https://www.mercedes-benz-bus.com/en_DE/brand/news/2023/kick-off-hanover-stationary-energy-storage-system-ecitaro-batteries.html.

Mercedes-Benz. (2016). World's largest 2nd-use battery storage is starting up. https://media.mercedes-benz.com/article/6a7e8349-4b24-479e-84f8-0d94f73406a4.

Mercedes-Benz. (2017). Daimler and enercity put battery replacement parts store for electric vehicles on the grid. https://media.mercedes-benz.com/article/ecb61007-028b-4d32-be0a-925afc5f74f8.

Mercedes-Benz. (2018). Shining example of the energy turnaround: Coal-fired power station becomes battery storage plant. https://media.mercedes-benz.com/article/41e52cc0-9760-43c3-9df2-b3f99eff784a.

Mercedes-Benz. (2024). Mercedes-Benz opens own recycling factory to close the battery loop. https://media.mercedes-benz.com/article/fe521181-3b57-4915-a51a-b5f6f352c574.

Miao, Y., Hynan, P., Von Jouanne, A., & Yokochi, A. (2019). Current Li-ion battery technologies in electric vehicles and opportunities for advancements. *Energies, 12*(6), 1–20. https://doi.org/10.3390/en12061074

Mobilize. (2022). Mobilize and Morbihan Energies join forces to increase the use of solar energy on Belle-Ile-en-Mer. https://media.mobilize.com/mobilize-and-morbihan-energies-join-forces-to-increase-the-use-of-solar-energy-on-belle-ile-en-mer-62668/?lang=eng.

Mossali, E., Picone, N., Gentilini, L., Rodrìguez, O., Pérez, J. M., & Colledani, M. (2020). Lithium-ion batteries towards circular economy: A literature review of opportunities and issues of recycling treatments. *Journal of Environmental Management, 264*, 110500. https://doi.org/10.1016/j.jenvman.2020.110500

Motus-E. (2023). Il riciclo delle batterie dei veicoli elettrici @2050: scenari evolutivi e tecnologie abilitanti. https://www.motus-e.org/wp-content/uploads/2023/03/Motus-E_PwCS_PoliMi_Il-riciclo-delle-batterie-dei-veicoli-elettrici-@2050-scenari-evolutivi-e-tecnologie-abilitanti.pdf.

Muller, P., Duboc, R., & Malefant, E. (2021). Recycling electric vehicle batteries: Ecological transformation and preserving resources. *Field Actions Science Report, Special Issue 23*, 74–81. http://journals.openedition.org/factsreports/6690.

Neumann, J., Petranikova, M., Meeus, M., Gamarra, J. D., Younesi, R., Winter, M., et al. (2022). Recycling of Lithium-ion batteries—current state of the art, circular economy, and next generation recycling. *Advanced Energy Materials, 12*(17). https://doi.org/10.1002/aenm.202102917.

Nickelhütte Aue. (2023). Batteries as a raw material. https://nha-aue.de/en/news/batteries-as-a-raw-material.

Northvolt. (2024). Sodium-ion cells. https://northvolt.com/products/cells/sodium-ion/.

Ordoñez, J., Gago, E. J., & Girard, A. (2016). Processes and technologies for the recycling and recovery of spent Lithium-ion batteries. *Renewable and Sustainable Energy Reviews, 60*, 195–205. https://doi.org/10.1016/j.rser.2015.12.363

POSCO. (2022). POSCO builds EV battery recycling plant in Poland. https://newsroom.posco.com/en/posco-builds-ev-battery-recycling-plant-in-poland/.

POSCO. (n.d.). Lithium/Nickel-what we do. Retrieved March 29, 2024, from https://www.posco-inc.com:4453/poscoinc/v3/eng/business/s91e2000400c.jsp.

Primobius. (n.d.-a). Locations. Retrieved April 3, 2024, from https://www.primobius.com/en-it/about-us/locations.

Primobius. (n.d.-b). Process & Technology. Retrieved March 29, 2024, from https://www.primobius.com/en-it/technology-services/recycling-process.

Ramoni, M. O., & Zhang, H.-C. (2013). End-of-Life (EOL) issues and options for electric vehicle batteries. *Clean Technologies and Environmental Policy, 15*(6), 881–891. https://doi.org/10.1007/s10098-013-0588-4

REDUX Recycling. (2024). REDUX recycling. https://www.redux-recycling.com/en/.

REDUX Recycling. (2018). Saubermacher opens new high-tech recycling plant for lithium-ion batteries. https://www.redux-recycling.com/wp-content/uploads/2018/06/PressRelease_Saubermacher-Redux-LithiumIonenBatterieRecycling_EN-1.pdf.

Renault Group. (2021). Renault eWays: The group presents two major new energy storage projects. https://media.renaultgroup.com/renault-eways-the-group-presents-two-major-new-energy-storage-projects/.

RWE. (2021). Second life for EV batteries: RWE and Audi create novel energy storage system in Herdecke. https://www.rwe.com/en/press/rwe-generation/2021-12-28-second-life-for-ev-batteries/.

SNAM. (2023). An episode dedicated to SNAM in Hyundai's original "Local Inspirations" program. https://www.snam.com/en/an-episode-dedicated-to-snam-in-hyundais-original-local-inspirations-program/.

SNAM. (n.d.). Recycling. Retrieved March 28, 2024, from https://www.snam.com/en/recycling-2/#etape.

Stena Recycling. (n.d.). Battery collection, recycling and reuse. Retrieved March 29, 2024, from https://www.stenarecycling.com/what-we-offer/material-recycling/batteries/.

SungEel HiTech. (n.d.). Recycling. Retrieved March 29, 2024, from https://www.sungeelht.com/en/html/12.

TES. (n.d.). Sustainable battery solutions. Retrieved March 28, 2024, from https://www.tes-amm.com/it-services/commercial-battery-recycling.
The Mobility House. (2024). Germany's largest multi-use storage facility opens on the EUREF campus. https://www.mobilityhouse.com/int_en/our-company/references/article/euref-campus-berlin.
Umicore. (2022). Umicore introduces new generation Li-ion battery recycling technologies and announces award with ACC. https://brs.umicore.com/en/news/umicore-introduces-new-generation-li-ion-battery-recycling-technologies-and-announces-award-with-acc/.
Umicore. (2023). Umicore battery recycling: Capturing profitable growth and enabling a circular and low-carbon battery value chain. https://www.umicore.com/en/newsroom/umicore-battery-recycling/.
VDI Zentrum Ressourceneffizienz. (2023). How to recycle lithium-ion batteries?–Closing the loop in e-mobility. https://www.youtube.com/watch?v=g1Ij4Emz8XQ.
Veolia. (2024). Electric car battery recycling. https://www.veolia.com/en/pollution/hazardous-waste/recycling-electric-car-batteries.
Volkswagen. (2021). Transforming old into new: Volkswagen group components commences battery recycling. https://www.volkswagengroup.it/en/media/press-releases/transforming-old-into-new-volkswagen-group-components-commences-battery-recycling.
Winslow, K. M., Laux, S. J., & Townsend, T. G. (2018). A review on the growing concern and potential management strategies of waste lithium-ion batteries. *Resources, Conservation and Recycling, 129*(July 2017), 263–277. https://doi.org/10.1016/j.resconrec.2017.11.001.
Zeng, X., Li, J., & Singh, N. (2014). Recycling of spent Lithium-ion battery: A critical review. *Critical Reviews in Environmental Science and Technology, 44*(10), 1129–1165. https://doi.org/10.1080/10643389.2013.763578
Zhang, W., Xu, C., He, W., Li, G., & Huang, J. (2018). A review on management of spent lithium ion batteries and strategy for resource recycling of all components from them. *Waste Management & Research, 36*(2), 99–112. https://doi.org/10.1177/0734242X17744655
Zhao, Y., Pohl, O., Bhatt, A. I., Collis, G. E., Mahon, P. J., Rüther, T., et al. (2021). A review on battery market trends, second-life reuse, and recycling. *Sustainable Chemistry, 2*(1), 167–205. https://doi.org/10.3390/suschem2010011

Electric Motors

Stefano Puricelli🄳 and Sebastián Fajardo Turner

Abstract As the transition to electric vehicles (EVs) accelerates globally, addressing the lifecycle of key components, such as electric motors, becomes increasingly relevant. Electric motors, particularly those based on permanent magnets (PM), contain valuable materials like copper, aluminium, and rare earth elements (REEs) such as neodymium and dysprosium. The recovery of these elements is essential for a sustainable resource management and for reducing the dependency on virgin materials. Despite mature end-of-life vehicle (ELV) frameworks in the European Union, challenges remain in effectively recycling electric motors, particularly in recovering REEs from PM motors. Current practices primarily involve destructive disassembly, which complicates the separation of REEs, often resulting in significant material losses. Innovative solutions are emerging, including mechanical, chemical, and hydrometallurgical methods for magnets recovery, alongside the development of alternative motor designs that minimise the reliance on critical materials. This chapter reviews the legislative landscape, technological advancements, and industrial initiatives addressing the recycling and remanufacturing of EV motors. By exploring case studies and European projects, it highlights practical approaches to enhance circularity within the EV sector.

Keywords Electric motor · Recycling · Permanent magnet · Neodymium · Rare earth elements

---

S. Puricelli (✉) · S. F. Turner
Department of Civil and Environmental Engineering, Politecnico di Milano, Milano, Italy
e-mail: stefano.puricelli@polimi.it

S. F. Turner
e-mail: sebastian.fajardo@mail.polimi.it

S. Puricelli
MatER Study Center, Laboratory for Energy and the Environment Piacenza, Piacenza, Italy

© The Author(s), under exclusive license to Springer Nature Switzerland AG 2025
M. Grosso and L. Rigamonti (eds.), *Waste Flows Generated by the Energy Transition*,
PoliMI SpringerBriefs, https://doi.org/10.1007/978-3-031-88951-6_3

1 Introduction

In the modern landscape of transportation, the passenger car remains the most widespread mode of transport, especially in developed nations. Within the European Union, cars are integral to daily life, accounting for two-thirds of daily commutes (Ortar & Ryghaug, 2019). Despite initiatives promoting public transport and cleaner modes of transport like walking and cycling, cars still lie at the centre of complex economic and social systems. This reliance, combined with the need to reduce carbon emissions and improve air quality within population centres, highlights the need for innovative propulsion systems, where Electric Vehicles (EVs) emerge as a promising solution. However, transitioning from traditional internal combustion engines to electric motors raises questions about resource availability and end-of-life management capabilities.

This transition is already underway, evidenced by the growing penetration of Plug-in Hybrids (PHEVs) and Battery Electric Vehicles (BEVs) in the most important markets, such as Europe, China, and the US. The evolution has been striking in Europe, where from a mere 600 new electric car registrations in 2010, the number leaped to approximately 1,061,000 in 2020, representing 11% of all new car registrations. The year 2021 witnessed an even more significant surge, with electricity-charged vehicles comprising nearly 18% of all newly registered passenger cars, equally split between BEVs and PHEVs (EEA, 2023). Globally, the electric vehicle population has expanded dramatically, reaching over 26 million in 2022—a 60% increase from 2021 and quintupling since 2018. This surge is primarily fuelled by the rapid sales growth in China, home to over half of the world's EVs (IEA, 2023).

Looking ahead, the European Commission aims to achieve tailpipe emission-free transportation for both passenger and commercial vehicles by 2050. Achieving this ambitious goal requires a significant reduction in CO_2 emissions from new vehicles, positioning electric vehicles to potentially dominate the transportation sector. This anticipated dominance calls for a marked increase in the demand for essential construction materials, particularly those critical for batteries, motors, electronics, and lightweight structures. Key among these materials are lithium, cobalt, and nickel—vital for lithium-ion batteries—and rare-earth elements (REE) such as neodymium, samarium, and dysprosium, essential for electric drive motors (Lipman & Maier, 2021). Furthermore, paired with an increased demand for materials, a nascent demand is expected for the management of the end of life (EoL) of the diverse components of EVs.

2 Types of Electric Motors

There are some key components within electric vehicles that distinguish them from the traditional internal combustion engine vehicle, particularly regarding EoL management considerations. EVs use a sophisticated array of power electronics, consisting of complicated networks of copper wiring, computerized electronics, and direct current (DC) to alternating current (AC) power inverters. For example, the Tesla Model S uses around 1.5 kms of copper wiring, and earlier editions used to employ upwards to 3 kms of wiring. Additionally, the electronic devices in the overall motor controller system require materials such as copper, silicon, plastics, steel, and small amounts of rarer materials such as gallium (Lipman & Maier, 2021). However, most evidently, the propulsion system is a key distinguisher, where EVs can use a varying array of electric motors with critical materials such as copper and REEs. In particular, neodymium (Nd), praseodymium (Pr) and dysprosium (Dy) are REEs used to make NdFeB magnets alloys for electric motors (Carrara et al., 2023). Nd is a key element of this kind of magnet, while Pr and Dy are used to improve the magnetic and physicochemical properties of the magnet (Li et al., 2024).

An electric motor consists of stator, rotor, bearings, housing, and shields (Fig. 1) (Carrara et al., 2023). Although there exists a wide array of technologies for electric motors, the current EV market is mainly dominated by Permanent Magnet (PM) motors and Induction Motors (IM) to some extent. Wound Rotor Synchronous Motors (WRSM) have started being used as an alternative that does not rely on permanent magnets and suffers from smaller losses compared to induction motors. In addition, the Synchronous Reluctance Motor (SRM) has been proposed as a possible alternative; however, SRMs are still not used in commercially available vehicles. In the following sections the workings, advantages and disadvantages, and a comparison of material compositions of these motors will be discussed. However, it is important to state that there are several variations in the technology of each family of motor, which may overlap with other types.

2.1 Induction Motors

An AC induction motor operates on the principles of electromagnetic induction. The motor consists of two main parts: a stator and a rotor. The stator, which is stationary, has coils that are supplied with an alternating current to produce a rotating magnetic field. This rotating field induces an electric current in the rotor, which is made of conductive bars embedded in a laminated iron core. The interaction between the magnetic field generated by the stator and the induced current in the rotor generates electromagnetic torque, causing the rotor to spin (Kirtley & Ghai, 1998). Unlike some other types of electric motors, the rotor in an AC Induction Motor does not receive any electrical power directly; instead, it is induced by the stator's magnetic field, hence its name. This design leads to a robust and relatively maintenance-free motor, making AC Induction Motors a popular choice in many industrial applications (de Santiago et al., 2012).

Induction motors, favoured for their simplicity and durability, offer significant benefits in EVs. They are known for their low maintenance needs, attributed to the absence of brushes and commutators, and their compatibility with regenerative braking systems. These motors are cost-effective to manufacture, efficient under

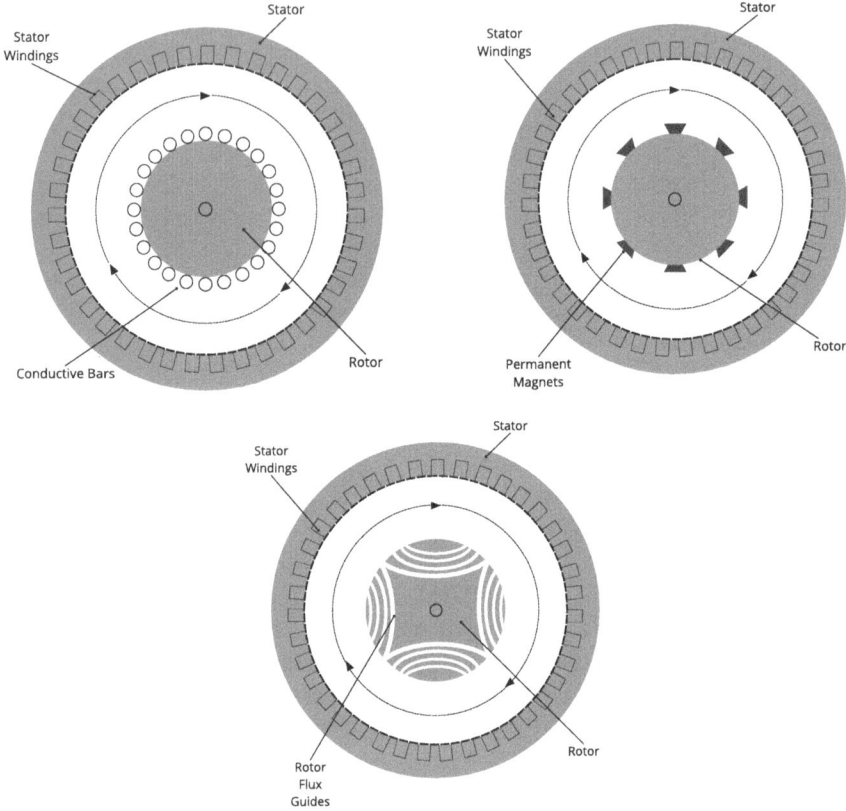

Fig. 1 On the top left: basic schematic of an induction motor, adapted from Kirtley and Ghai (1998). On the top-right: basic schematic of a permanent magnet motor, adapted from Kirtley (1998). On the bottom: basic schematic of a synchronous reluctance motor, adapted from Ayad Alkhafaju and Uzun (2018)

variable loads, and reliable in harsh conditions, making them suitable for automotive use. Additionally, they present safety advantages due to a lower risk of overheating (Yildirim et al., 2014).

However, induction motors also have drawbacks. They typically exhibit lower torque at low speeds and require complex control systems for precise speed and torque management. Compared to other motor types like permanent magnet motors, they can be bulkier and heavier for the same power output. The peak efficiency of induction motors may be lower, and they depend on AC power inverters, adding to the vehicle's powertrain complexity. Finally, they experience significant losses of energy in the rotor (Yildirim et al., 2014).

As an alternative to induction motors, wound-rotor synchronous motors (WRSM) have started being used in the EV industry. Also known as an externally excited synchronous motors (EESMs), the WRSMs replace the magnets on the rotor with coil windings that can be supplied with a direct current to generate a magnetic field. This design allows for independent control of both the stator and rotor fields, offering enhanced flexibility in motor performance. However, this comes with the downside of requiring additional manufacturing steps to incorporate windings into the rotor and the need for brushes to transmit power to the rotor. Historically, these motors have had lower power and torque density compared to PM motors, but modern advancements have brought their performance to a comparable level. Renault has been an early adopter of this technology in their Zoe model, and it has since been embraced by other major manufacturers like BMW and Nissan (IDTechEx, 2023). Possible advantages of these motors are the absence of permanent magnets, and reduced losses compared to traditional induction motors.

2.2 Permanent Magnet Motors

Permanent magnet motors also consist of a stator and rotor like induction motors. However, PM motors utilise permanent magnets embedded in the rotor to create a magnetic field, unlike the induction motor, which relies on electromagnetic induction in the rotor (Kirtley & Ghai, 1998). This fundamental difference leads to PM motors offering higher efficiency and torque at low speeds, but at a higher cost and complexity, compared to the simpler and more robust design of induction motors (Yildirim et al., 2014). Similarly to induction motors, the interaction between the magnetic field generated by the stator and the magnetic field in the rotor produces electromagnetic torque, allowing it to spin.

There are two main types of configurations for the permanent magnets. They can be mounted in the surface of the rotor, referred to as Surface Permanent Magnets (SPM), or they can be located within pockets close to the rotor, known as Integrated Permanent Magnets (IPM).

PM motors have gained popularity in the EV market thanks to the high energy density they can provide, their high efficiency due to the lack of rotor losses, and their excellent performance at lower speeds. However, they do present a variety of drawbacks. On the technical side, they suffer from high sensitivity to temperature changes, which requires sophisticated thermal management systems. On the logistic side, they require permanent magnets which are often composed of REEs, such as neodymium, which are considered critical raw materials and fetch high price in the market and supply risks (Yildirim et al., 2014).

2.3 Synchronous Reluctance Motors

Synchronous Reluctance Motors (SRMs) have been proposed as a possible alternative to both IM and PM motors; they mainly differ in the fact that they generate torque purely through the magnetic reluctance of their uniquely designed rotor, without relying on induced currents or permanent magnets. SRMs consist of a stator, like that of an induction motor, equipped with copper windings, and a unique rotor that is designed without any windings or permanent magnets. The rotor is typically made of laminated steel, featuring a series of slots or barriers to facilitate magnetic flux paths. These components work together, with the stator's rotating magnetic field inducing a magnetic response in the rotor, aligning the rotor's reluctance paths with the stator's magnetic field to produce torque (Kirtley & Ghai, 1998).

This construction offers benefits such as uncomplicated and sturdy construction, fault-tolerant operation, straightforward control, and impressive torque-speed characteristics. SRM drives can inherently operate within an extensive constant-power range. Nevertheless, they exhibit drawbacks, including torque ripple and acoustic noise, which, while not prohibitive, require consideration in EV applications. They are gaining attention due to concerns about material costs in mass EV production (Hashemnia & Asaei, 2008).

3 Material Composition Considerations

In July 2023, the European Commission published a material composition breakdown of end-of-life passenger cars after de-pollution. Using data composition of passenger cars from ELVs in the EU according to Eurostat, the European Commission calculated a standard passenger car weight and composition for Internal Combustion Engine Vehicles, Hybrid Vehicles, Plug-in Hybrid Vehicles, and Battery Electric Vehicles (Table 1 and Fig. 2). The baseline assumes the total weight as equal for all types of vehicles.

These numbers include the motor weight, however oversimplified, since materials used in small quantities such as in the electric motors or power electronics are not reported. It is important to clarify that this is a general approximation that aims at characterising an enormous diversity of passenger vehicles with a single number. It is expected that the real total weight of each type of vehicle will vary.

There is no clear standard or readily available information on specific material compositions by weight for the electric motors used in EVs. It is important to state that EV electric motors tend to have a much higher energy density than those used in the industrial sectors. A review done by Elwert et al. (2015) approximated the weight of an 80-kW PM motor in a BEV at 62.5 kg and that of a 20-kW PM motor in a hybrid electric vehicle (HEV) at 44.0 kg. Tazi et al. (2023) reported an average weight of 48.8 kg for a PM-free motor in BEVs, while, regarding PM motors, they reported average motor weights of 44.9 kg (in BEVs), 34.5 kg (in PHEVs), and 21.3 kg (in HEVs). Typical weight composition of electric motors for EVs are shown in Table 2.

As previously stated, there exists a significant lack of information on the material composition of electric motors, and no consistent information for IMs and SRMs can be found. However, from Table 2 it can be observed that the materials in PM motors are mostly metallic. Regarding the REEs in the PMs, Elwert et al. (2015) estimate 30% of the NdFeB magnets is constituted by REEs, signifying approximately

Table 1 Material composition of end-of-life vehicles (passenger cars) after de-pollution. Data adapted from European Commission (2023a)

Material	Internal combustion engine vehicle (kg)	Hybrid vehicle (kg)	Plug-in hybrid vehicle (kg)	Battery electric vehicle (kg)
Steel	653	660	621	642
Cast iron	101	101	96	16
Wrought aluminium	40	58	76	108
Cast aluminium	79	91	93	77
Copper	14	20	23	35
Magnesium	5	5	5	1
Manganese	8	8	8	7
Glass	24	21	22	26
Average plastic	159	129	143	166
Rubber	41	34	38	39
Glass-fibre reinforced plastic	9	4	5	5
Others	5	6	7	14
Total	1,137	1,137	1,137	1,137

0.63 kg and 0.42 kg in the 80-kW motor and 20-kW motor, respectively (Elwert et al., 2015). When applying the same REE content to the data by Tazi et al. (2023), a BEV PM motor would contain 0.38 kg of REEs. Nonetheless, these results present a contradiction with current European Commission estimates, which approximate 1.2 kg of REEs used in PM motors per EV (European Commission, 2023a).

A study done by Ballinger et al. (2019) identified that in 2018, twenty models were responsible for 51% of the EV sales. Of those, all PHEVs and 62% of BEVs used electric motors relying on REEs. Assuming these models as representative data of the full market, between 75 and 86% of all EVs sold required REEs (Ballinger et al., 2019). However, given the considerable expansion in the EV market and the rapid development of technologies, a new study might be necessary to understand the material flows in the EV market. An added degree of complexity when estimating the material flows in electric vehicles in general is the fact that EVs can have a significant variation in number of motors and their placement. EVs may have one motor with or without a transmission, two motors (one for each axle), or even four motors (one per wheel) (Xue et al., 2009).

Furthermore, the European Commission identifies other types of EVs which may use different configurations of electric motors. These are broken down in Table 3.

In terms of raw materials demand, following the study of Carrara et al. (2023), Table 4 is compiled, highlighting also the specific use of the identified material for the electric motors.

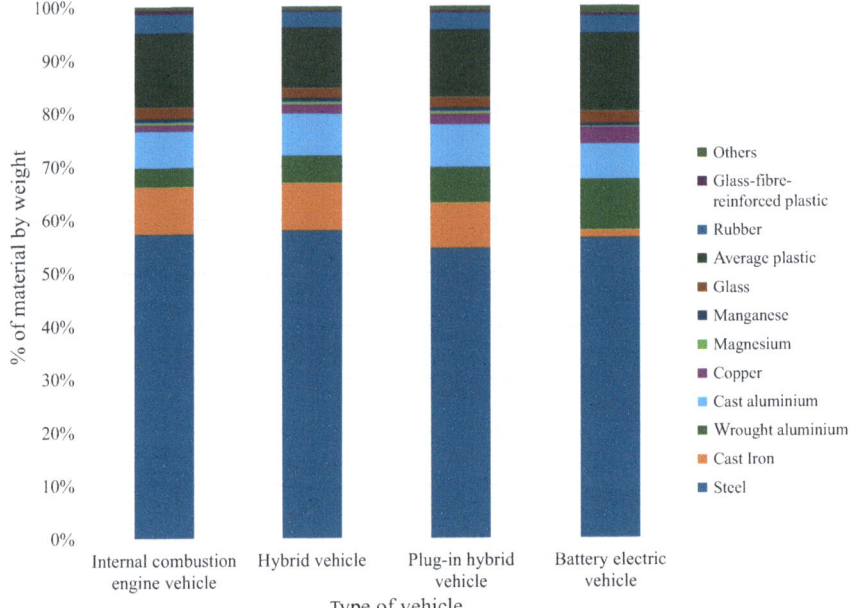

Fig. 2 Material composition of end-of-life vehicles (passenger cars) after de-pollution. Data based on European Commission (2023a)

Table 2 Bill of main materials (% m/m) of electric motors for EVs. Data from Tazi et al. (2023)

Material	PM motors (%)	PM-free motors (%)
Aluminium	32.1	29.5
Copper	10.9	20.6
Electrosteel	50.2	46.2
Insulation	2.1	1.9
Magnets	2.8	–
Steel	1.6	1.5
Other	0.4	0.4

4 Legislative Framework

4.1 Directive on End of Life for Vehicles

In the European Union, a mature end of life management for vehicles already exists, as governed by the Directive 2000/53/EC on end-of-life vehicles (European Parliament & Council of the EU, 2000). The Directive sets out measures to prevent and limit waste from end-of-life vehicles (ELVs) and their components, by ensuring their reuse, recycling, and recovery. It also aims to improve the environmental performance

Table 3 Motor types in non-passenger car vehicles. Data adapted from European Commission (2023a)

Vehicle	Motor type	Main critical raw materials
Electric two-wheeler (EU brands)	Permanent Magnet Synchronous Motors (PMSM) with a typical weight of 19 kg	Laminated Si-steel and REE in PMs
Electric two-wheeler (non-EU brands)	Brushless Direct Current PM motor, with a typical weight range from 4 to 15 kg	Laminated Si-steel and REE in PMs
Electric lorries	More than 93% use PMSMs and PM assisted reluctance motors	Laminated Si-steel and REE in PMs
Electric buses	More than 99% of BEVs and PHEVs use PMSMs	Laminated Si-steel and REE in PMs

Table 4 Selection of raw materials employed for electric motors (Carrara et al., 2023). *Note* Critical raw materials are underlined

Use	Material
Electric motor (without permanent magnets)	• Aluminium in casings and as lightweight material in other metal parts • Copper in windings, cables, inverters, control systems • Chromium in stainless steel and alloys resistant to corrosion in various motor components • Iron as cast iron or in steel composition for casings • Molybdenum in stainless steel and alloys for electric motor's housing and shaft • Silicon in semiconductors and control electronics, and as alloying element in Al-alloys and silicon steel
Permanent magnets	• Boron • Dysprosium as additive in NdFeB magnets • Iron • Neodymium in NdFeB magnets in the motor's rotor for providing strong magnetic field • Praseodymium together with neodymium • Silicon in semiconductors and control electronics, and as alloying element in Al-alloys and silicon steel

of all economic operators involved in the life cycle of the vehicles. The key targets to be attained are two: reuse and recovery of EoL vehicles shall be at least 95%, while reuse and recycling shall be at least 85%. In 2022, in the EU, the reuse/recovery and reuse/recycling rates were, respectively, 94.4% and 89.1% (Eurostat, 2024). The same targets, this time referred to new vehicles, were set by the Directive 2005/64/EC (European Parliament & Council of the EU, 2005). In brief, this Directive imposes that a new vehicle (cars, station wagons, people carriers and light commercial vehicles) is manufactured to be reused/recovered to a minimum of 95% (m/m) and reused/recycled to a minimum of 85% (m/m) and.

4.2 On-Going Review on ELV Directive

The original ELV Directive was published on the 18th of September 2000. In the time since, new opportunities and threats have been identified for the EoL of vehicles, which led to the launch of a review on the Directive in 2021. The review found an assortment of key problems in the automotive industry. A lack of circularity in design and production can still be noted as the existing laws have not led to better eco-designs of vehicles nor to an increased use of recycled materials. In general, a poor quality of vehicle waste treatment was identified, in particular low-quality scrap steel, low plastics recycling rate and poor separation of materials. Furthermore, there is a high dependency on imported raw materials. The production of vehicles is one of the most resource-intensive industries in EU: it is the top consumer of aluminium (42%), magnesium (44%), platinum group metals (63%), natural rubber (67%), and REEs (estimated to be 30% by 2025, and growing exponentially mainly due to increased EV sales) (European Commission, 2023b).

Due to these key issues, amongst others, a new proposal for the Directive was released in July 2023. Aligned with the European Green Deal and the Circular Economy Action Plan, the ELV Regulation proposal is founded upon and supersedes two prevailing Directives: Directive 2000/53/EC concerning end-of-life vehicles and Directive 2005/64/EC pertaining to the type-approval of motor vehicles in terms of their reusability, recyclability, and recoverability (European Commission, n.d.-b).

The proposed new rules address various stages of a vehicle lifecycle, from its design and market placement to its final treatment at the end of life. The objectives include improving the circular design of vehicles to facilitate the removal of materials for reuse and recycling. The proposal also aims to ensure that a minimum of 25% of the plastic used in vehicle construction comes from recycling, of which 25% sourced from recycled end-of-life vehicles. Additionally, there is a focus on recovering more and higher-quality raw materials, including critical raw materials, plastics, steel, and aluminium.

In order to ensure adequate financing for the mandatory ELV treatment, producers have to be held financially responsible for vehicles once they become waste. This will also encourage recyclers to improve the quality of their processes. Furthermore, the rules seek to expand coverage to include additional vehicle categories such as motorcycles, lorries, and buses, gradually extending EU regulations to ensure proper end-of-life treatment across the entire automotive spectrum (European Commission, n.d.-b).

The Directive review identified the electrification trend of the vehicular fleet as a key challenge and area of opportunity. Increased electrification will cause a significant increase in the demand for critical materials such as copper, aluminium, REEs, and others. Now the production and EoL stages gain a heightened importance in the life cycle of the vehicles, as compared to the use phase. Thus, ensuring the recovery of CRMs in the automotive sector is an essential element for the future of energy transition and the overall EU strategy to improve the security of supply of such materials (European Commission, 2023a). Concerning REEs, the Directive aims for 350 t of rare earths in PM materials to be separately collected for reuse and recycling in 2035, and 1,500 t in 2040, significantly contributing to the EU's efforts for strategic autonomy.

However, the Directive review also identified key challenges in achieving such recovery of materials from ELVs. Since vehicles do not currently abide by specific legal requirements on their design or end-of-life phases, circularity or design for recycling are not necessarily integrated as a requirement in the design of EVs. Furthermore, there exists a significant lack of information on CRMs content and location in these vehicles, which could prevent EU dismantlers and recyclers from properly recovering these materials from collected vehicles. Finally, dismantlers are not currently experienced with such components and the markets for them are not yet developed (European Commission, 2023a).

Additionally, the Directive review identified the need for a new generation of electric drive systems to enable a large-scale adoption of EVs. Such a new generation of electric drivetrains would need to reduce dependency on REEs, while improving energy efficiency, power density, and reducing manufacturing/recycling costs. Advanced Reluctance Motors were proposed as a possible solution to this challenge (European Commission, 2023a).

5 End-of-Life Management Practices in a Circular Approach

The general recycling process for ELVs, both internal combustion and electric, involves several key steps, i.e., de-pollution, dismantling, shredding, and post-shredding sorting (Fig. 3). It is important to note that, according to the law, consumers can turn in ELVs without incurring in any cost.

The initial phase of the recycling process entails de-pollution. Here, the goal is to remove potentially hazardous components and toxic materials such as batteries, exhaust oil, oil filter, refrigerant, fuel, airbags, lamps, monitors, and others (Fondazione Per lo Sviluppo Sostenibile & AIRA, 2022). Then, specific parts are dismantled, either due to their potential for resale (e.g., engines and spare parts) as well as their potential for recovery (e.g., plastics, glass, tires), or because they could compromise the shredding phase or contaminate the shredder residues (Fondazione Per lo Sviluppo Sostenibile & AIRA, 2022).

The following dismantling phase includes the removal of catalyst, metallic components, tyres, big plastic components, dashboard, tanks, and glass components (Fondazione Per lo Sviluppo Sostenibile & AIRA, 2022). The removal of catalysts,

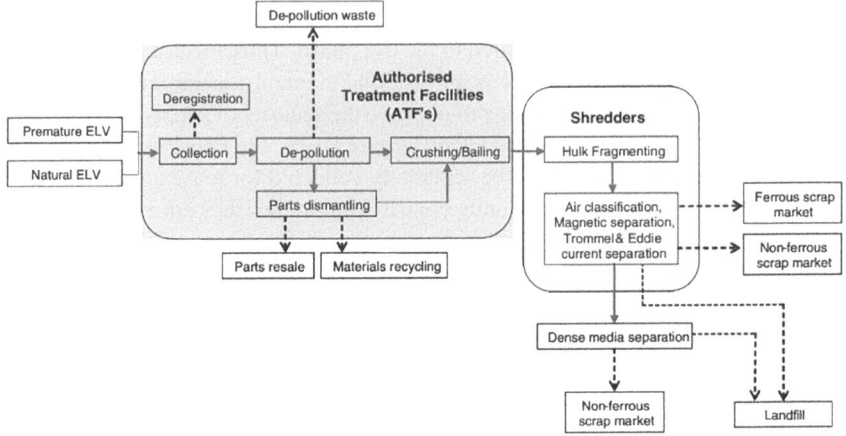

Fig. 3 The flow of the vehicle through EoL operations. Reprinted from Edwards et al. (2006)

tyres, and metal components are the most diffuse operations, because profitable. The most suitable European List of Waste (ELW) code for waste electric motors is, to the authors' knowledge, 16 02 16 ("components removed from discarded equipment other than those mentioned in 16 02 15"[1]).

After dismantling, the ELVs undergo shredding (Sakai et al., 2014). During this stage, the vehicles are broken down typically through a hammer mill and processed using various techniques, including sieving, magnetic separators (to recover iron concentrate), eddy-current separators (for mixed non-iron metal concentrate), and density separators (to separate plastic, dust, and lightweight materials). The residue generated from the auto shredding process, known as ASR (Automotive Shredded Residue), primarily consists of light components, metals, and glass-rich shredder sand.

Under Directive 2000/53/EC, the removal of batteries before the shredding and post-shredding processes is mandated. However, the removal of electric motors and power electronics is not explicitly prescribed by the law; nonetheless, it is considered a necessary step for any attempt to recover minor metals. Once removed and disassembled, Induction Motors and Synchronous Reluctance Motors (as well as any motor not relying on Permanent Magnets) can be relatively easily recycled using the conventional routes for iron, steel, aluminium, and copper. In the case of PM motors, most of its weight also consists of these conventional metals, however, the recovery of REEs within the permanent magnets poses different challenges.

Usually, combustion engines are removed from the end-of-life vehicles before the shredding phase (Elwert et al., 2017). The same process will happen for EVs, because copper and REEs can be recycled more efficiently instead of being disseminated over the metallic fractions. Recycling of conventional electric motors is already state of the art. It could not be otherwise, since motors consume around 40% of the electrical

---

[1] 16 02 15: hazardous components removed from discarded equipment.

energy globally generated (Tiwari et al., 2021). Although a non-destructive disassembly would better preserve the sub-components of the motor, this practice tends to be largely manual and time-consuming. The current recycling industry uses destructive disassembly (e.g., shredding) (Li et al., 2024). A series of mechanical treatments allow to separate the easy-to-recover copper, steel, and aluminium. However, the magnets are stuck to process units, lost in the magnetic ferrous fraction, and diluted in the other products (Li et al., 2024). Currently, REE recovery from PM motors is null (Tazi et al., 2023).

A remanufacturing process, instead, includes the steps of collection, primary inspection, disassembly, secondary inspection, repairing, testing, and final assembly (Tiwari et al., 2021). Common damaged elements are the bearings, which must be replaced, and the stator windings, which can be rewound (Tiwari et al., 2021). An example of disassembly is shown in Fig. 4.

In the next sub-sections, the recycling of magnets from PM motors is deepened. In fact, PM motors are the most usual type of electric motor in EVs (Carrara et al., 2023; IDTechEx, 2023) and the recovery of REEs is a future challenge that recyclers of electric motors will need to face.

5.1 Disassembly

The research project MORE (Elwert et al., 2015) explored a range of methods to disassemble the components found within an electric motor, from manual labour to heat and chemical processes, as well as mechanical methods (Table 5). However, the high labour costs associated with manual disassembly in countries with high wages made it an impractical choice. The use of heat resulted in the magnets breaking, and the chemical method was both resource-intensive and energy-consuming. As a result, the project decided to focus on mechanised and automated dismantling methods as a more cost-effective and efficient approach for disassembling and recycling permanent magnets from end-of-life components.

For IPMs, rotor-specific ejectors were designed to press out the magnets from their rotor segments. On the other hand, SPM rotors need to be freed of the bandages before magnet extraction. Then, the magnets are shorn off with a nonmagnetic wedge and transported to a storage chamber. In summary, the MORE project proved that it is possible to use a mechanised dismantling method. However, further tests on varying designs of motors would be needed (Elwert et al., 2017).

5.2 Reuse of Magnets

Theoretically, the direct reuse of the magnetic components would be the most beneficial option from the economic and environmental points of view. However, this poses several challenges. A possible degradation of the magnetic properties could occur

Fig. 4 Disassembly step for a 3-kW motor. Reprinted from Tiwari et al. (2021)

Table 5 Magnet disassembly methods explored by "MORE" Project. Data adapted from Elwert et al. (2017)

Method	Notes
Manual extraction	Could prove to be the most exact, however, due to high labour-costs in developed countries, this approach was discarded
Thermal destruction of adhesive bonds including demagnetisation	Led to violent and erratic separation of magnets in SPMs Caused magnet fractures due to thermal expansion
Chemical detachment	Required high amounts of chemicals and energy, thus discarded
Automated mechanical separation	Proved the most convenient Different methods were developed for SPMs and IPMs Had difficulties in rotors with grooves May prove difficult with a large variety of motors

whether during the use phase or the EoL processing, for example, ruptures during the extraction of brittle magnets and degradation and accumulation of impurities, such as glue residues (Elwert et al., 2017). Furthermore, there is the possibility that there would be no demand for the reuse of these pieces in new generations vehicles, due to constant innovation in magnets and electric motors.

5.3 Remanufacturing of Magnets

In case reuse is not a viable solution, the magnets could be remanufactured onto other magnets without fully going back to its raw materials. To do this, it is essential to have magnet scraps of known and homogeneous composition. Thorough removal of impurities, such as glue residues and coatings, is crucial, as even impurity concentrations in the ppm range can deteriorate the magnetic properties. The carbon and oxygen content after cleaning should not exceed 100 and 300 ppm compared to the magnet alloy (Elwert et al., 2017). The primary challenge in reprocessing magnets from electric drive motors is removing glue residues. The MORE project experimented with both chemical (e.g., hydrolysis with caustic soda or dimethylformamide) and mechanical (e.g., grinding with SiC-based granules) techniques. Grinding proved effective in removing glue residues, but magnets with notches or grooves posed challenges (Elwert et al., 2017).

Three remanufacturing methods for the magnetic alloy were investigated by MORE: direct remelting in a vacuum induction furnace, purely mechanical comminution, and comminution after hydrogen decrepitation (Table 6). Direct remelting proved problematic due to the formation of slag and poor separability of the said slag and the desired alloys, so it was not further pursued. Pretreatment through hydrogen decrepitation led to better grindability than with purely mechanical methods, thus the

Table 6 Reprocessing methods explored by "MORE" Project. Data adapted from Elwert et al. (2017)

Method	Notes
Direct remelting in vacuum induction furnace	Slag formation due to oxygen content Poor separation of alloy and slag Low efficiency
Purely mechanical comminution	Crushers and rollers were used for comminution before jet milling
Comminution after hydrogen decrepitation	Hydrogen decrepitation was performed before the mechanical processes Better grindability

study continued using this process. Reactions between hydrogen and magnets form interstitial hydrides and provoke a volumetric expansion that makes the magnets friable (Li et al., 2024; Sabbe Moosa, 2016).

A 3% loss in remanence (the ability to retain magnetisation) was observed in magnets with recycled content compared to magnets made entirely from primary raw materials. The remanence loss was linearly related to the recycling material content: it is stated that a 1% remanence loss is expected for every 10% of recycled content. Additionally, repeated reprocessing could result in downcycling due to non-magnetic oxide and carbide formation (Elwert et al., 2017). Additionally, corrosion resistance was tested, and magnets with 30% recycling material had an increased weight loss compared to those made from primary raw materials, though still within acceptable limits. However, magnets with higher recycling material amounts could exceed acceptable corrosion limits (Elwert et al., 2017).

While reprocessing NdFeB magnets is feasible, using magnets containing recycled material may not be realistic for electric drive motors in (H)EVs due to the need for high remanence. Nonetheless, it could be used in products with less strict demands. Finally, it is to be considered that the MORE experiments were performed with production scraps, which contains fewer impurities than cleaned end-of-life scraps (Elwert et al., 2017).

5.4 Raw Material Recovery from Magnets

For the case of raw material recovery, it is important to consider that high magnetic properties must be preserved, thus impurities play a critical role. Additionally, any process for raw material recovery must also be able to handle metallic and oxidised NdFeB wastes, because oxidised grinding sludge constitutes a major waste stream besides production and end-of-life scrap (Elwert et al., 2017). Several approaches and processes have been developed. These methods can be broadly classified into three main categories: gas-phase extraction, pyrometallurgical methods, and hydrometallurgical methods. Each of these methods has its own unique set of advantages and disadvantages. Other studied methods are electrochemical and bio-metallurgical (Li et al., 2024).

Gas-phase extraction, also known as gas-phase reduction, is a method used in the recycling of permanent magnets, particularly for recovering rare-earth elements. In this process, the crushed magnet material is heated in a controlled atmosphere, typically using hydrogen gas. At elevated temperatures, the rare-earth elements in the magnet powder react with the hydrogen, forming gaseous compounds, such as hydrides. These hydrides can then be easily separated from other non-valuable materials in the mixture. Subsequently, the hydrides can be decomposed to release the rare-earth elements, which are then purified and used to manufacture new magnets. However, a notable disadvantage of gas-phase extraction is that it requires high-temperature processing, which can be energy-intensive and costly. Special precautions must be taken to ensure safety due to the use of flammable hydrogen gas (Binnemans et al., 2013).

In the pyrometallurgical process, crushed magnet materials are subjected to high temperatures, typically in the range of 1,000–2,000 °C, in a controlled environment, such as a furnace or smelter. Examples of pyrometallurgical processes are roasting, molten extraction methods, and carbon-hydrolysis (Li et al., 2024). At these elevated temperatures, the magnet material undergoes thermal decomposition, releasing the REEs which can then be collected, separated from the non-REE constituents, and further processed into new magnets (Binnemans et al., 2013). One advantage of pyrometallurgy is its efficiency in separating rare-earth elements from other materials, since the high temperature can cause the magnet components to melt and segregate based on their melting points. Additionally, pyrometallurgy requires less water and fewer chemical reagents compared to hydrometallurgy, reducing costs and hazardous waste production (Li et al., 2024). However, there are also notable disadvantages, such as the high energy consumption associated with the elevated temperatures, leading to increased environmental impact and operational costs. Moreover, pyrometallurgy may not be suitable for magnets containing elements that are sensitive to high temperatures or prone to oxidation (Binnemans et al., 2013).

In the hydrometallurgical method, the magnets are first crushed into a fine powder, and then the powdered material is dissolved in an acidic solution, for example hydrochloric or sulphuric acid. This acid leaching process selectively dissolves the rare-earth metals, leaving behind other non-valuable materials. After leaching, the solution is subjected to various chemical and physical processes, such as precipitation, solvent extraction, ionic liquids extraction, membrane technologies, to separate and purify the rare-earth elements. Finally, these purified elements can be used to manufacture new magnets, thus enabling the sustainable recycling of valuable materials while reducing the need for mining and resource extraction. Nonetheless, hydrometallurgy for rare-earth magnet recycling has drawbacks, including its multi-step process, which can be time-consuming, the high demand for chemicals, and the generation of significant quantities of wastewater, making it resource-intensive and environmentally challenging (Binnemans et al., 2013).

In the specific context of the MORE project, the hydrometallurgical approach was identified as the most suitable option. This is primarily because it offers adaptability to address different impurities and can effectively handle both metallic and oxidised waste materials. It provides the project with the versatility needed to achieve its goals (Elwert et al., 2017).

5.5 Performance, Benefits, Drawbacks

Strong points and disadvantages of the recycling methods for magnets from PM motors are summarised in Table 7.

Table 7 Summary of possible recycling methods for magnets from PM motors. Adapted from Binnemans et al. (2013)

Method	Advantages	Disadvantages
Direct re-use in current form/shape	• Most economical way • Low input of energy • No consumption of chemicals • No waste generated	• Only for easily accessible magnets (a challenge for EVs) • May not be able to be directly reused due to rapid innovation and change in the EV market
Reprocessing of alloys to magnets after hydrogen decrepitation	• Less energy input required than for hydro- or pyrometallurgical routes • No waste generated	• Not applicable to mixed scrap feed • Not applicable to oxidised magnets • Worse properties compared to a new magnet
Gas-phase extraction	• Generally applicable to all types of magnets • Applicable to oxidized and unoxidized alloys • No generation of wastewater	• High-temperature processing, which is energy-intensive and costly • Special precautions needed to prevent leaks of gasses
Pyrometallurgical methods	• Generally applicable to all types of magnet compositions • No wastewater generated • Fewer steps required compared to hydrometallurgical methods • REEs can be obtained in metallic state	• Large energy requirements • Oxidised magnets are incompatible with direct smelting and liquid metal extraction • Electroslag refining and the glass slag method generate large amounts of solid waste
Hydrometallurgical methods	• Generally applicable to all types of magnet compositions • Applicable to oxidised and non-oxidised alloys • Same processing steps as those used in REE extraction from primary ores • Active attempts to improve through Green Chemistry (Bandara et al., 2016)	• Many steps are necessary to have a new magnet • High chemical demand • Generation of large amounts of wastewater

There is a lack of LCA studies regarding electric motors recycling (Li et al., 2024). However, LCA literature reports that NdFeB magnets reused or made with recycled materials have lower climate change impacts than new NdFeB (Jin et al., 2018; Li et al., 2024). The recycling of magnets could also be an economic opportunity for decreasing the costs of electric motors, whose economic value is made up for 40–60% by REEs (Li et al., 2024).

6 Industrial Initiatives

A thorough EoL management and material recovery for EVs implies a large network of companies and organisations that may carry on diverse tasks. Information about recycling plants for EoL electric motors in Europe are scarce. This can signify that the export of electric motors to non-European countries (e.g., China) is predominant.

As previously stated, there is no fully implemented process in the Western World to dismantle and recover REEs from electric motors. Nonetheless, the European Union began the funding of "REEPRODUCE" project, whose goal is *"to set up the first sustainable and complete European REEs-recycling value chain at industrial scale, able to produce REEs from EoL products at competitive cost and with environmentally friendly technologies"* (REEPRODUCE, n.d.).

24 European initiatives related to the recycling of electric motors, especially the permanent magnets ones, are enlisted in Table 8.

7 Conclusions

Electric motors (EMs) for electric vehicles (EVs) consist of stator, rotor, bearings, housing, and shields. The permanent magnet motors are the most widespread type of EM utilised in EVs. EMs are mainly made of metals and contain critical and strategic raw materials such as aluminium, copper, and silicon; permanent magnets contain other key raw materials like boron and rare earth elements (dysprosium, neodymium, and praseodymium).

In the EU, the management of end-of-life vehicles is governed by the Directive 2000/53/EC, currently under revision. This regulation set a minimum target of 95% reuse and recovery of vehicles and 85% reuse and recycling of vehicles. In the EU the two rates were, respectively, 94.4% and 89.1% in 2022.

Although the dismantled motors can be easily recycled using the conventional routes for metal scraps, the recovery of rare earth elements from permanent magnets EMs poses different challenges. The current recycling industry uses a destructive disassembly (e.g., shredding) to ease the recovery of copper, steel, and aluminium. However, the magnets are stuck to process units or diluted and lost in the magnetic ferrous fractions. Remanufacturing and recycling of permanent magnets are currently under study, with 24 industrial and research initiatives currently underway in Europe.

Table 8 European industrial and research initiatives related to electric motors

Group	Technology	Location	Highlights	Sources
Bronneberg	Mechanical	Netherlands	Their motor wrecker is a machine consisting of three stations: breaking of the housing and removal of the rotor; splitting of the stator in two parts; separation of iron and copper of the stator. In the REEPRODUCE project, Bronneberg will develop a machine or machine-line which can separate the magnets	Bronneberg (n.d.), REEPRODUCE (n.d.)
Ceit	Mechanical	Spain	Focuses on the use of robotics, artificial vision, deep learning and reinforced learning to develop a pilot plant for the automated dismantling of PM motors from WEEEs	REEPRODUCE (n.d.)
ECOMAGNET	Mechanical	Spain	Production of recycled anisotropic NdFeB magnet powder which, after mixing with epoxy resin, can be then used as raw material for manufacturing new magnets	ECOMAGNET (n.d.)
Elkem	High-temperature electrolysis	Norway	In the REEPRODUCE project, Elkem will receives REE-oxalates for calcination to produce Rare Earth Oxide mixture, used in Rare Earth Alloy production	REEPRODUCE (n.d.)
Euregio Recycling	Mechanical	Belgium, Netherlands	A company specialised in electric motors recycling. In a first plant, materials are shredded. In a second plant, shredded metals are converted into clean products	Euregio Recycling (n.d.), REEPRODUCE (n.d.)
GlobEco	Mechanical	Italy	In the INSPIREE project, this company will implement a plant that will dismantle 1,000 t/y of electric rotors to recover mag-nets	European Commission (n.d.-a), Itelyum (n.d.)

(continued)

Electric Motors

Table 8 (continued)

Group	Technology	Location	Highlights	Sources
HyProMag	Mechanical and hydrogen decrepitation	Germany, UK	Through its patented technology, NdFeB magnets are broken by hydrogen at ambient conditions. Ni coating can be separated mechanically	HyProMag (n.d.), Maginito (n.d.)
Indumetal Recycling	Mechanical	Spain	Company that recycles WEEEs and complex scrap. In the REEPRODUCE project, Indumetal Recycling will coordinate the development of an automated sorting and dismantling system, will supply Nd-based PMs from (WEEEs and will host a pilot plant for PM extraction from HHDs	REEPRODUCE (n.d.)
Inovertis	Hydrometallurgy and high temperature electrolysis	France	In the REEPRODUCE project, it oversees the design, the engineering, the construction and the commissioning of the hydro-metallurgical as well as of the high temperature electrolysis pilots. It has also to perform the social, environmental and economic evaluation of the rare earth recycling chain in comparison to the current scenario	REEPRODUCE (n.d.)
Institute FAPS of the Friedrich-Alexander-Universität Erlangen-Nürnberg (FAU)	Separation technologies	Germany	In the REEPRODUCE project, it conducts research in two scenarios: optimisation of process chains to enable automated dis-assembly of magnets from rotors; methods and technologies to recover magnet materials after shredding complete rotors	REEPRODUCE (n.d.)
Itelyum	Hydrometallurgy	Italy	In the INSPIREE project, Itelyum will deal with the construction and operation of a plant that will receive 2,000 t/y of PMs and, through hydrometallurgy, will recover 700 t/y of REE oxalates	European Commission (n.d.-a), Itelyum (n.d.)

(continued)

Table 8 (continued)

Group	Technology	Location	Highlights	Sources
ITR Recycling Systems	Mechanical	Italy	It designs recycling plants for electric motors, allowing to re-cover copper, aluminium, and iron. The recycling process includes hammer mills, magnetic separators, and eddy current separators	ITR Recycling Systems (n.d.)
JGI-HYDROMETAL	Hydrometallurgy	Belgium	In the REEPRODUCE project, JGI-HYDROMETAL will: operate and validate pilot technology to recover REEs from EoL Nd-based PMs and obtain REEs-oxalates; lead upstream pilot extraction of Nd-based PM from sorted EoL products; lead downstream pilot conversion of REEs-oxalates into REE-alloy; test production of new PMs using recycled REEs	REEPRODUCE (n.d.)
Magneti Ljubljana	Permanent magnet production	Slovenia	In the REEPRODUCE project, it will produce PMs from recycled metals and alloys made by project partners	REEPRODUCE (n.d.)
MIMplus Technologies	Permanent magnet production	Germany	This company is developing a process capable to make PMs from recycled NdFeB magnets. The process consists of hydrogen decrepitation, milling, feedstock production, injection moulding of the base material and alignment of the magnetic field, binder removal, and sintering step with magnetisation	MIMplus Technologies (n.d.)

(continued)

Electric Motors 67

Table 8 (continued)

Group	Technology	Location	Highlights	Sources
Panizzolo	Mechanical	Italy	It designs plants and machinery for various recycling processes, including electric motors. The recycling machineries for electric motors include shredders, hammer mills, magnetic drum separators, x-ray separators, sieves, eddy current separators, zig-zag air density separators, and others	Panizzolo (n.d.)
The Remet Company	Mechanical	UK	Breakage of electric motors into Cu, Fe, and Al fractions	The Remet Company (n.d.)
RarEarth	Chemical	Italy	Startup dedicated to the recycling of magnets coming from end-of-life electric motors of two-wheelers. The proposed project aims to chemically recycle NdFeB magnets	RarEarth (n.d.)
Revac	Mechanical	Norway, Sweden	In the REEPRODUCE project, REVAC will contribute with sampling and knowledge about Nd in WEEEs	REEPRODUCE (n.d.)
SINTEF	High temperature electrolysis	Norway	In the REEPRODUCE project, SINTEF will develop the high temperature electrolysis pilot to produce rare earth alloys for permanent magnets	REEPRODUCE (n.d.)
Sense2sort—Toratecnica	Sensor based separation	Spain	In the REEPRODUCE project, Sense2sort will work in the development of a semi-mobile sorting machine for the separation of EoL WEEE-devices containing Neodymium PMs	REEPRODUCE (n.d.)
Stena Recycling	Mechanical	Sweden	An agreement between ABB and Stena Recycling offers a recycling option to ABBs costumers with regards to large electric motors and generators over 10 t. The same concept exists for small motors	ABB (2022), Stena Recycling (2019)

(continued)

Table 8 (continued)

Group	Technology	Location	Highlights	Sources
STOKKERMILL		Italy	It develops machinery and recycling systems for electric motors. The recycling line is generally a combination of hammer mill, delamination mill, magnetic separator, eddy current separator, densimetric table, and optical separator	STOKKERMILL (n.d.)
TECNALIA	Hydrometallurgy	Spain	In the REEPRODUCE project, Tecnalia will focus on the hydrometallurgical technology, by designing and studying adequate pre-treatments, leaching systems, metal separation and recovery routes	REEPRODUCE (n.d.)

Information about specific end of life treatment plants for EMs in Europe is scarce; this can signify that the export of electric motors to non-European countries is predominant.

Competing Interests The authors have no conflicts of interest to declare that are relevant to the content of this chapter.

References

ABB. (2022). Partnering together for a circular economy: ABB large motors and generators sweden and stena recycling. https://new.abb.com/news/detail/90905/partnering-together-for-a-circular-economy-abb-large-motors-and-generators-sweden-and-stena-recycling.

Ayad Alkhafaji, M., & Uzun, Y. (2018). Design and analysis of synchronous reluctance motor (SynRM) using MATLAB Simulink. In *Proceedings of the international conference on innovative research in science engineering & technology*. https://doi.org/10.33422/irset.2018.12.32.

Ballinger, B., Stringer, M., Schmeda-Lopez, D. R., Kefford, B., Parkinson, B., Greig, C., et al. (2019). The vulnerability of electric vehicle deployment to critical mineral supply. *Applied Energy, 255*, 113844. https://doi.org/10.1016/j.apenergy.2019.113844.

Bandara, H. M. D., Field, K. D., & Emmert, M. H. (2016). Rare earth recovery from End-of-Life motors employing green chemistry design principles. *Green Chemistry, 18*(3), 753–759. https://doi.org/10.1039/C5GC01255D.

Binnemans, K., Jones, P. T., Blanpain, B., Van Gerven, T., Yang, Y., Walton, A., et al. (2013). Recycling of rare earths: A critical review. *Journal of Cleaner Production, 51*, 1–22. https://doi.org/10.1016/j.jclepro.2012.12.037.

Bronneberg. (n.d.). Electric motor recycling. Retrieved May 28, 2024, from https://www.bronneberg-recycling.co.uk/solutions/electric-motor-recycling/.

Carrara, S., Bobba, S., Blagoeva, D., Dias, A., Cavalli, P., Georgitzikis, A., et al. (2023). *Supply chain analysis and material demand forecast in strategic technologies and sectors in the EU-a foresight study*. JRC Science for Policy Report. EU CRM. https://doi.org/10.2760/334074.

European Commission. (2023a). SWD(2023) 256 final. https://eur-lex.europa.eu/legal-content/EN/TXT/?uri=SWD:2023:256:FIN.

de Santiago, J., Bernhoff, H., Ekergård, B., Eriksson, S., Ferhatovic, S., Waters, R., et al. (2012). Electrical motor drivelines in commercial all electric vehicles: A review. *IEEE Transactions on Vehicular Technology, 61*(2), 475–484. https://doi.org/10.1109/TVT.2011.2177873.

ECOMAGNET. (n.d.). ECOMAGNET. Retrieved June 5, 2024, from https://ecomagnet.es/en/home-en/.

Edwards, C., Coates, G., Leaney, P. G., & Rahimifard, S. (2006). Implications of the End-of-Life vehicles directive on the vehicle recovery sector. *Proceedings of the Institution of Mechanical Engineers, Part B: Journal of Engineering Manufacture, 220*(7), 1211–1216. https://doi.org/10.1243/09544054JEM473SC.

EEA. (2023). New registrations of electric vehicles in Europe. https://www.eea.europa.eu/en/analysis/indicators/new-registrations-of-electric-vehicles?activeAccordion=.

Elwert, T., Goldmann, D., Römer, F., Buchert, M., Merz, C., Schueler, D., et al. (2015). Current developments and challenges in the recycling of key components of (Hybrid) electric vehicles. *Recycling, 1*(1), 25–60. https://doi.org/10.3390/recycling1010025.

Elwert, T., Goldmann, D., Roemer, F., & Schwarz, S. (2017). Recycling of NdFeB magnets from electric drive motors of (Hybrid) electric vehicles. *Journal of Sustainable Metallurgy, 3*(1), 108–121. https://doi.org/10.1007/s40831-016-0085-1.

Euregio Recycling. (n.d.). Euregio recycling. Retrieved May 29, 2024, from https://www.euregio-recycling.com/en/.
European Commission. (n.d.-a). LIFE22-ENV-IT-INSPIREE. Retrieved June 5, 2024, from https://webgate.ec.europa.eu/life/publicWebsite/project/LIFE22-ENV-IT-INSPIREE-101113882/industrial-production-of-mixed-rare-earth-elements-oxides-and-carbonates-from-spent-magnets-recycling.
European Commission. (n.d.-b). End-of-Life vehicles. Retrieved March 5, 2024, from https://environment.ec.europa.eu/topics/waste-and-recycling/end-life-vehicles_en.
European Commission. (2023b). SWD(2023) 257 final. https://eur-lex.europa.eu/legal-content/EN/TXT/?uri=SWD%3A2023%3A257%3AFIN&qid=1689323637403.
European Parliament and Council of the EU. (2000). Directive 2000/53/EC. *Official Journal of the European Communities, L 269*. https://eur-lex.europa.eu/legal-content/EN/TXT/?uri=celex%3A32000L0053.
European Parliament and Council of the EU. (2005). Directive 2005/64/EC. *Official Journal of the European Union, L 310*. https://eur-lex.europa.eu/legal-content/IT/ALL/?uri=CELEX%3A32005L0064.
Eurostat. (2024). End-of-Life vehicle statistics. https://ec.europa.eu/eurostat/statistics-explained/index.php?title=End-of-life_vehicle_statistics.
Fondazione Per lo Sviluppo Sostenibile, & AIRA. (2022). Studio sulle problematiche del riciclo e recupero dei veicoli fuori uso. https://www.fondazionesvilupposostenibile.org/wp-content/uploads/dlm_uploads/Studio-sulle-problematiche-del-riciclo-e-recupero-dei-veicoli-fuori_Susdef-AIRA-2022_.pdf.
Hashemnia, N., & Asaei, B. (2008). Comparative study of using different electric motors in the electric vehicles. In *2008 18th International Conference on Electrical Machines* (pp. 1–5). https://doi.org/10.1109/ICELMACH.2008.4800157.
HyProMag. (n.d.). Rare earth magnet recycling. Retrieved June 5, 2024, from https://hypromag.com/rare-earth-magnet-recycling/.
IDTechEx. (2023). 4 Ways to Eliminate Rare Earths in EV Motors and One You Haven't Heard. https://www.idtechex.com/en/research-article/4-ways-to-eliminate-rare-earths-in-ev-motors-and-one-you-havent-heard/29723.
IEA. (2023). Global EV outlook 2023. https://www.iea.org/reports/global-ev-outlook-2023.
Itelyum. (n.d.). The INSPIREE project. Retrieved June 5, 2024, from https://www.itelyum.com/en/inspiree-2/.
ITR Recycling Systems. (n.d.). Impianti di trattamento motori elettrici. https://www.itrimpianti.com/en/.
Jin, H., Afiuny, P., Dove, S., Furlan, G., Zakotnik, M., Yih, Y., et al. (2018). Life cycle assessment of neodymium-iron-boron magnet-to-magnet recycling for electric vehicle motors. *Environmental Science and Technology, 52*(6), 3796–3802. https://doi.org/10.1021/acs.est.7b05442.
Kirtley, J., & Ghai, N. (1998). Induction motors. In J. L. Kirtley, Jr., H. W. Beaty, N. K. Ghai, S. B. Leeb, & R. H. Lyon (Eds.), *Electric motor handbook* (1st ed.). McGraw-Hill Education. https://www.accessengineeringlibrary.com/content/book/9780070359710/chapter/chapter4.
Kirtley, J. (1998). Permanent magnet-synchronous (Brushless) motors. In J. L. Kirtley, Jr., H. W. Beaty, N. K. Ghai, S. B. Leeb, & R. H. Lyon (Eds.), *Electric motor handbook* (1st ed.). McGraw-Hill Education. https://www.accessengineeringlibrary.com/content/book/9780070359710/chapter/chapter6.
Li, Z., Hamidi, A. S., Yan, Z., Sattar, A., Hazra, S., Soulard, J., et al. (2024). A circular economy approach for recycling electric motors in the End-of-Life vehicles: A literature review. *Resources, Conservation and Recycling, 205*, 107582. https://doi.org/10.1016/j.resconrec.2024.107582.
Lipman, T. E., & Maier, P. (2021). Advanced materials supply considerations for electric vehicle applications. *MRS Bulletin, 46*(12), 1164–1175. https://doi.org/10.1557/s43577-022-00263-z.
Maginito Limited. (n.d.). Maginito. Retrieved June 5, 2024, from https://maginito.com/.

MIMplus Technologies. (n.d.). Research & Development. Retrieved June 3, 2024, from https://magnets.mimplus.com/.

Ortar, N., & Ryghaug, M. (2019). Should all cars be electric by 2025? The electric car debate in Europe. *Sustainability (Switzerland), 11*(7). https://doi.org/10.3390/su11071868.

Panizzolo. (n.d.). Impianti di riciclaggio per motori elettrici. https://www.panizzolo.com/riciclaggio-metalli/impianti-di-riciclaggio/impianto-di-riciclaggio-per-motori-elettrici/.

RarEarth. (n.d.). RarEarth. https://www.rarearth.it/.

REEPRODUCE. (n.d.). Work plan. REEPRODUCE. Retrieved March 7, 2024, from https://www.reeproduce.eu/work-plan/.

Sabbe Moosa, I. (2016). History and applications in the production of permanent magnets. *International Journal of Advanced Research in Engineering and Technology, 7*(5), 37–44. http://iaeme.com/Home/journal/IJARET37editor@iaeme.com, http://iaeme.com/Home/issue/IJARET?Volume=7&Issue=5, http://iaeme.com.

Sakai, S., Yoshida, H., Hiratsuka, J., Vandecasteele, C., Kohlmeyer, R., Rotter, V. S., et al. (2014). An international comparative study of End-of-Life vehicle (ELV) recycling systems. *Journal of Material Cycles and Waste Management, 16*(1), 1–20. https://doi.org/10.1007/s10163-013-0173-2.

Stena Recycling. (2019). Sustainable collaboration with ABB. https://www.stenarecycling.com/news-insights/newsroom/2019/an-abb-and-stena-recycling-collaboration-towards-a-more-sustainable-industry/.

STOKKERMILL. (n.d.). Linea di riciclaggio motori elettrici. https://www.eu.stokkermill.com/recycling-plants-and-lines/linea-riciclaggio-motori-elettrici.

Tazi, N., Orefice, M., Marmy, C., Baron, Y., Ljunggren, M., Wäger, P., et al. (2023). Initial analysis of selected measures to improve the circularity of critical raw materials and other materials in passenger cars. *Publications Office of the European Union*. https://doi.org/10.2760/937462.

The Remet Company. (n.d.). Electric motors recycling . Retrieved May 14, 2024, from https://www.remetcompany.com/scrap-metals/motors/.

Tiwari, D., Miscandlon, J., Tiwari, A., & Jewell, G. W. (2021). A review of circular economy research for electric motors and the role of industry 4.0 technologies. *Sustainability (Switzerland), 13*(17), 9668. https://doi.org/10.3390/su13179668.

Xue, X. D., Cheng, K. W. E., & Cheung, N. C. (2009). Selection of electric motor drives for electric vehicles. In *2008 Australasian Universities Power Engineering Conference*. https://ieeexplore.ieee.org/document/4813059.

Yildirim, M., Polat, M., & Kurum, H. (2014). A survey on comparison of electric motor types and drives used for electric vehicles. In *2014 16th International Power Electronics and Motion Control Conference and Exposition* (pp. 218–223). https://doi.org/10.1109/EPEPEMC.2014.6980715.

Wind Turbine Blades

Gaia Brussa

Abstract Wind energy is a cornerstone of Europe's renewable energy transition, with installed capacity significantly increased over the past two decades and many turbines now aging and nearing the end of their operational life. While most of a wind turbine is considered recyclable, composite materials used in blades remain an issue, due to their complex characteristics and durability, making recycling challenging. Innovative recycling methods are being developed to improve the recovery of both reinforcing fibres and resins. Mechanical recycling, already at the industrial scale (TRL 9), is prevalent but yields lower-quality materials with limited secondary applications. Thermal methods, such as pyrolysis and fluidized bed combustion, provide higher recovery rates but are energy intensive. Chemical recycling, particularly solvolysis, are promising but are still at the laboratory scale and expensive, making only carbon fibre recovery an economically viable application. Cement co-processing is identified as a valid short-term solution, integrating blade materials into cement production and reducing the demand for raw materials and fuels. Exploring the European legislative landscape, a lack of unified regulations for composite waste is revealed. Industry advocates for specific waste codes, landfill bans, and harmonized guidelines to enhance recycling efforts.

Keywords Energy transition · Waste · Renewable energy · Recycling · Wind turbine blade · Composite material

1 Wind Energy

Although the lifespan of a wind turbine is relatively long, reaching more than 20 years on average (Khalid et al., 2023), at some point a significant number of windmills are going to reach the end of their useful life and it will be necessary to plan their

---

G. Brussa (✉)
Department of Civil and Environmental Engineering, Politecnico di Milano, Milano, Italy
e-mail: gaia.brussa@polimi.it

MatER Study Center, Laboratory for Energy and the Environment Piacenza, Piacenza, Italy

© The Author(s), under exclusive license to Springer Nature Switzerland AG 2025
M. Grosso and L. Rigamonti (eds.), *Waste Flows Generated by the Energy Transition*,
PoliMI SpringerBriefs, https://doi.org/10.1007/978-3-031-88951-6_4

management as waste, in order to recover the maximum value from their dismantling. For example, WindEurope (2024) provided an estimation of the average age of onshore wind farms in Europe (Fig. 1) showing that a significant amount of wind turbines (in terms of installed capacity) is approaching the end of life.

It can be observed that Germany, Spain, and Italy have a significant number of wind turbines over 20 years old. Germany, in particular, has about 10 GW of capacity nearing the end of its service life. These three countries also have the highest average turbine ages, all exceeding 10 years. Additionally, other countries have turbines in the 15–19 year range, indicating that many of these turbines will reach the end of their operational life in the coming years. Moreover, the cumulative installed capacity, both onshore and offshore, is projected to nearly double by 2030 compared to 2023 from about 200 GW to almost 400 GW (Fig. 2).

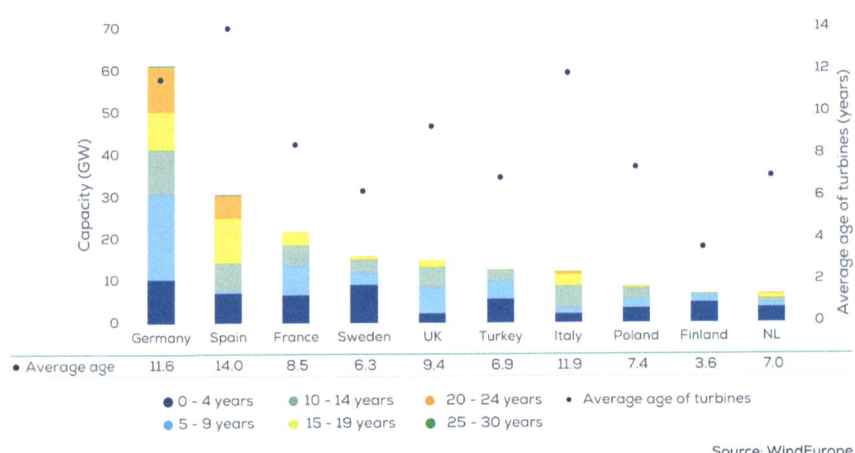

Fig. 1 Average age of onshore wind farms in Europe (WindEurope, 2024)

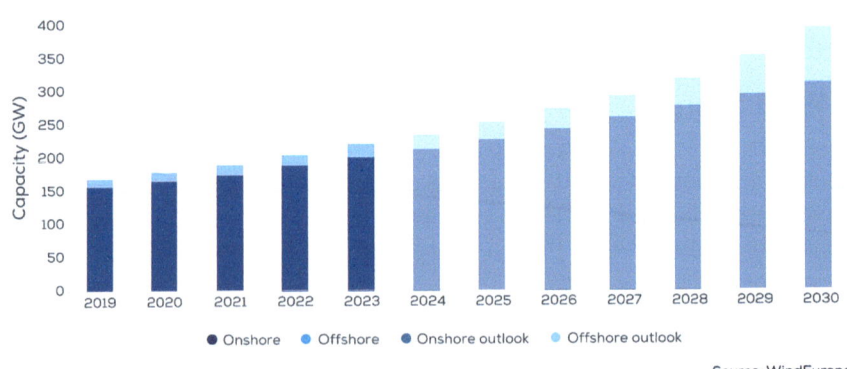

Fig. 2 WindEurope's outlook on wind power capacity in the EU (WindEurope, 2024)

Several studies have addressed the issue of estimating the amount of waste from blades expected to be generated in Europe in the coming years (DecomBlades, n.d.; Lichtenegger et al., 2020; Sommer et al., 2020). However, the forecasts show significant variability and seldom provide detailed country-level data.

2 Characterisation of the Waste Flow

Wind turbines consist of a tower, i.e., a large tubular steel section erected on a foundation, a nacelle, which contains many of the electrical and mechanical components, and a rotor, comprising blades, a hub and a blade pitch system. A wind power plant also includes other components such as the high voltage transformer, switchgears and electrical cables (Carrara et al., 2020). Excluding the foundation, the tower represents the most significant element of a wind turbine, reaching 60% of the turbine's total mass, while the nacelle and rotor account for about 20% each. Table 1 reports the average material composition of a wind turbine (Fioretti et al., 2023).

It is commonly assumed that around 85–90% of a windmill is recyclable, since many parts are made of metals, such as steel and aluminium, or copper in wires and electronics; therefore, there are well-established recycling processes and markets exist for the resulting secondary materials (WindEurope, 2021a). The most relevant open point is the fate of the non-metallic portion and especially the composite materials used in the rotor blades. In fact, extensive research efforts have been made and are ongoing to improve the mechanical properties of wind turbines and to extend their service life. Especially blades must be designed to sustain extreme mechanical stresses and endure severe environmental conditions for decades. This results in the use of materials that pose significant challenges when they reach the end of life and need to be recycled (Rathore & Panwar, 2023).

The most common composite materials, which account for most of the blade weight, are fibre reinforced polymers i.e., glass-fibre reinforced plastic (GFRP) and carbon fibre reinforced plastic (CFRP): the first are the most common (~98%) while the latter can be found only on recent models (~2%) (Fioretti et al., 2023).

In general, composite materials consist of a discontinuous phase, usually fibrous (called reinforcement), dispersed within a continuous homogeneous phase (called matrix or resin). The fibres (typically glass, carbon, boron, or aramid fibres), thanks

Table 1 Average composition of a wind turbine by weight (Fioretti et al., 2023)

Material	Composition (weight) (%)
Steel and stainless steel	66–79
Iron or cast iron	5–17
Copper	1
Aluminium	0–2
Composite materials	10–15

to their high mechanical properties, provide rigidity and mechanical strength to the material, while the surrounding matrix (typically polymeric) transfers the stresses between the fibres and protects them. Composite materials combine the best characteristics of two or more components to generate a product with superior physical and mechanical properties compared to those of the individual constituent materials acting independently. The fibre reinforced plastics are commonly used due to their excellent mechanical, wear and corrosion resistance; they also show long lifespan at reasonably priced manufacturing costs, given by the low-cost materials (Jani et al., 2022; Khalid et al., 2023).

The composition of these materials can be approximated as 60% reinforcing fibres and 40% polymer matrix (Fonte & Xydis, 2021).

2.1 Fibre Reinforced Plastics

In 2023 the total European composites market reached a volume of 2,559 kt (AVK, 2024). Glass fibre reinforced plastic (GFRP) accounted for more than 95% of the total composites market (AVK, 2024). GFRP can vary in terms of matrix type, with the main groups being thermosetting plastics and thermoplastics, and in terms of fibre length, distinguishing between short fibres (less than 2 mm), long fibres (between 2 and 50 mm), or continuous fibres (more than 50 mm). Among fibreglass with thermosetting resin, there are also multi-axial fabrics (non-crimp-fabrics, NCF), which are fabrics consisting of several layers of glass fibre reinforced plastics stacked on top of each other and sewn together with a thin yarn using frames (Fioretti et al., 2023).

Thermoset materials such as polyester and epoxy resins are the most commonly used as matrices; the employment of thermoset resins confers high durability but also flexibility to the finished material. Thermoplastics are an alternative to thermosets, providing the advantage of recyclability (Mishnaevsky et al., 2017). As of today, thermoplastics cover almost 60% of the European composites market (AVK, 2024).

In 2023, the transportation sector emerged as the largest user of GFRP in Europe (about 50% of the total production). Other significant sectors include electrical and electronic equipment, with 21.2%, and the construction sector accounting for 18.5% of the total production volume (AVK, 2024). The application in construction encompasses the usage of composites not only in buildings and infrastructures but also in wind turbines (according to the classification provided by Colledani and Turri 2022). Currently, the wind energy sector represents a wide application area for GFRP and is expected to undergo significant growth in the near future. However, specific data regarding the use of GFRP in this sector are not available. The only identified data for GFRP use in the wind energy sector is on a global scale: in ORE Catapult (2016) it was estimated that as of 2021 about 2.5 million tonnes of composite material were globally in use in the wind energy sector, of which 12–15 t/MW of generated power are associated solely with the use of GFRP (Compositi Magazine, 2023). It is important to note that the current volume of such waste is expected to progressively increase in the coming years, not only due to the increasing use of these materials in various sectors but also due to the progressive accumulation of infrastructures, recently installed and that are going to reach the end of life after 20 years of operation (e.g., wind turbines). Regarding the wind energy sector, research shows that waste from blades will reach 30 kt/year in Europe by 2026, increasing to a value of 50 kt/year by 2030 (Colledani & Turri, 2022).

The main alternative to GFRP is carbon fibre reinforced plastics (CFRP). Carbon fibres exhibit superior mechanical properties compared to glass fibres, albeit at a higher cost (ranging from 4 to 40 times the price of glass fibres, depending on the specific application). Due to their specific properties, CFRPs are mostly used in high-value sectors such as wind energy, sports equipment, sports cars, construction, and the aerospace and aircraft industries (Colledani & Turri, 2022).

As of 2023, CFRP cover only a small portion of the total European composite market (2.5%). However, in recent years processing volumes of CFRP have been increasing in contrast to GFRP that showed a slight decrease (AVK, 2024).

2.2 Wind Turbine Blade Structure

Despite varying material compositions among blade types and blade manufacturers, it is estimated that fibreglass-reinforced materials constitute an average of 85% of the mass of a blade, followed by the core (9%), metal parts (3%), coatings and adhesives (3%) (Fioretti et al., 2023). Glass fibres generally are the primary reinforcement in wind turbine blades, while carbon fibres are used to a lesser extent. As polymer matrix materials thermosets are widely used, especially epoxies, polyesters, vinyl esters or polyurethane. The sandwich core is made of balsa wood or foams such as polyvinyl chloride (PVC) or polyethylene terephthalate (PET) (WindEurope et al., 2020). Polyurethane (PUR) can be used both as coating and structural adhesive; alternatives are unsaturated polyester resin (UPR) and epoxies, respectively. Figure 3 illustrates the typical structure of a wind turbine blade.

2.3 Material Requirements

The growing size of turbines contributes to the increase in the capacity factor, an indicator of the actual production of energy vs the nominal power, and to the reduction in material intensity. Taller towers, wider rotors and lighter drivetrains have enabled higher capacity factors, which have risen from an average of 27% in 2010 for newly commissioned onshore projects to 34% in 2018 (IRENA, 2019). These trends have also helped to reduce the material intensity for some materials. Frequently, a table with the mass of blade corresponding to the rated capacity of the wind turbine has been provided (Cooperman et al., 2021; Jani et al., 2022); however, the data (Table 2) stem from Liu and Barlow (Liu & Barlow, 2017) result now outdated, being based

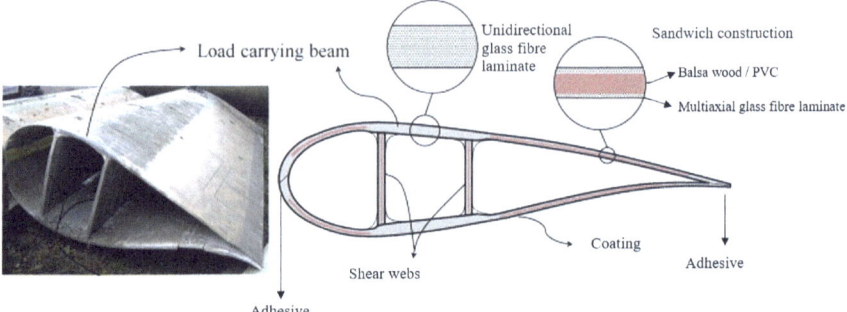

Fig. 3 Cross-section and general composition of a wind turbine blade. Reprinted from Beauson et al. (2016) with permission from Elsevier

Table 2 Average blade mass per unit rated power for single turbines (Liu & Barlow, 2017)

Rated capacity of wind turbine (MW)	Mass of blade (kg/kW)
<1	8.43
1–1.5	12.37
1.5–2	13.34
2–5	13.41
>5	12.58

on data obtained from 1998 to 2014 and due to the limitation to 5 MW as maximum rated power, which is no longer the case, with the most recent offshore turbines approaching 15–16 MW.

Recently, more detailed correlations between the mass of wind turbine blades and various characteristics of the wind turbine or plant (e.g., rated power, wind turbine or rotor mass, rotor diameter) have been developed to improve the estimation of the composite waste generated at the end of life. However, continuous technological advancements require constant efforts in updating these values and relationships. Sommer et al. (2020) provided a list of peer-reviewed articles addressing the issue as of 2020.

2.4 Critical Raw Materials in Wind Turbines

Wind turbines are largely cited when dealing with energy transition and raw materials demand: for example, in the study by Carrara et al. (2023) the connection to critical raw materials is highlighted. Critical raw materials (CRMs) use in wind turbines is mostly determined by steel, which requires manganese during its production and nickel and niobium in specifical alloys, and permanent magnets which are used in direct-drive generators. In fact, the main difference among the various wind turbines is due to the generators that can be classified as gearbox or direct-drive (DD). The two types have significantly different manufacturing, differing in generator design and drivetrain system. As a result, both the mass and the material content differ greatly between the two. Nevertheless, both configurations can include permanent magnets (Carrara et al., 2020).

Neodymium-iron-boron (NdFeB) permanent magnets are the state-of-the-art technology used in wind turbines: currently produced NdFeB permanent magnets contain on average 28.5% neodymium, and 4.4% dysprosium in weight. Moreover, since the rated power of turbines is increasing and magnets can weigh up to 4 t, their consumption is rapidly rising (Carrara et al., 2020).

Table 3 summarises a selection of raw materials employed in wind turbines according to their function and highlights the presence of critical and strategic raw materials.

Additionally, blades require to be designed to optimise energy production, which is proportional to their length, and to resists harsh environmental conditions and variable speed, all while minimising their weight. The key to achieve these objectives is using materials with a high strength-to-weight ratio combined with high stiffness and fatigue resistance. Balsa wood possesses these characteristics and therefore is applied in spar caps and blade cores. Despite not being classified as critical raw material, balsa wood can determine supply bottlenecks since approximately 90% of its production is concentrated in Ecuador (Carrara et al., 2023). To reduce this dependency, some alternatives are being considered by wind blade manufacturers, such as recycled polyethylene terephthalate (rPET) or innovative biocomposites based on hemp hurd cellulose (Carrara et al., 2023).

Table 3 Selection of raw materials used in wind turbines and their function Carrara et al. (2023). *Note* Critical and strategical raw materials are underlined

Function	Material
Structural function	Chromium and molybdenum for stainless steel
	Manganese for steel production
	Niobium microalloying element for high structural steel
	Silicon as alloying element in high-performance steel and in silicone for sealants, adhesives, lubricants
	Zinc for anti-corrosion coatings
Electrical connection	Copper and aluminium for wires
	Nickel in alloys and stainless steel
	Chromium for stainless steel
	Lead for soldering and submarine cable sheathing
Permanent magnets (generator)	Iron, boron and neodymium for NdFeB
	Dysprosium as additive in NdFeB
	Praseodymium in permanent magnets

3 Legislative Framework

So far, the legislation on the treatment of composite waste in general or wind turbine blades at the end of life is rather limited both at the EU and national levels.

Waste from GFRP can be split into three main groups: small components, such as sports and leisure equipment and various consumer goods, for which there is no separate collection at the domestic and commercial level, neither nationally nor in Europe; large infrastructures, transportation means and construction waste, including wind turbine blades, aircraft, boats, vehicles, and building structures; and waste generated during fibreglass production and processing, that can be directly collected from the plants and have a known composition.

Referring to the European List of Waste (ELW), there is no unique identification for fibreglass waste. Currently, the most used code is ELW 17.02.03, i.e., plastic waste from construction and demolition activities. Additionally, the following codes are often used nationally for the wind turbine sector (Fioretti et al., 2023):

- ELW 07.02.13: plastic waste (from organic chemical processes).
- ELW 10.11.03: waste glass-based fibrous materials from the manufacture of glass and glass products (waste from thermal processes).
- ELW 10.11.12: waste glass other than those mentioned in 10.11.11*[1] (waste from thermal processes).
- ELW 10.11.99: wastes not otherwise specified (waste from thermal processes).
- ELW 12.01.05: plastics shavings and turnings (waste from shaping and physical and mechanical surface treatment of metals and plastics).

It must be noted that currently these codes may be associated with GFRP waste from many other sectors than wind energy.

In May 2024, seven industry associations[2] related to composites-sector published a statement asking to improve the EU regulatory frameworks on composite materials and waste to actively improve their circularity. The first request is a revision of the ELW and the inclusion of dedicated waste codes for end-of-life composite materials. In fact, in their opinion, this would be the first step to clearly identify these streams, to assess the required infrastructure for collection, sorting and treatment and finally to develop business cases for their management. Moreover, they suggest including specific codes for decommissioned rotor blades and nacelles and for boats at the end of life, since they are identified as significant composite material waste streams. Specific waste codes could also promote the introduction of reuse and recycling targets (EuCIA, 2024).

---

[1] Waste glass in small particles and glass powder containing heavy metals.

[2] EuCIA (the European Composites Industry Association), Cembureau (European Cement Association), WindEurope (presenting wind sector), EBI (European Boating Industry), Glass Fibre Europe (Association of European continuous filament glass fibre industry), Epoxy Europe (Association of European epoxy resin producers) and UP/VE Resins (European Unsaturated Polyester and Epoxy Vinyl Ester Resin Association) as part of Cefic (European Chemical Industry Council).

By now, in the absence of precise data and a unique ELW code for identifying GFRP waste, the annual quantification of this type of waste is complex, and currently only estimates and approximations are available, both at the global, European, and national levels.

Another relevant topic by industry associations is the need for a landfill ban on composite waste. In fact, in 2021 WindEurope, as an association that gathers wind turbine manufacturers, component suppliers, power utilities and wind farm developers, financial institutions, research institutes and national wind energy associations, called for a Europe-wide landfill ban on decommissioned wind turbine blades by 2025 in order to accelerate the development of sustainable practices and of recycling technologies for blades' composite materials, which at the present pace are not expected to be fully deployable at scale before 2030 (WindEurope, 2021a). Recently, also EuCIA, Cembureau, EBI, Epoxy Europe, Glass Fibre Europe, UP/VE Resins have joined WindEurope in this request for dedicated policies (EuCIA, 2024). So far, only four EU countries (Germany, Austria, the Netherlands, and Finland) make a clear reference to composite waste in their national waste legislation, where landfilling or incineration of composites are forbidden (WindEurope et al., 2020).

In Germany the ban refers to the direct landfilling of waste having an organic content higher than 5% starting from 2009; the resins used in the blades generally are organic and therefore the regulation applies to this waste. As an alternative, waste of glass fibre reinforced polymers is sent to cement co-processing, typically applying a gate fee of around 150 €/t (WindEurope et al., 2020).

In Finland the landfilling ban is similar, but the organic content threshold is set to 10%; the most common waste management option, missing the recycling route, is energy recovery (Fioretti et al., 2023).

In the Netherlands, the landfilling of composite waste was banned by the National Waste Management Plan. Nevertheless, exemptions are provided in case the alternative treatment cost for wind farm operators is higher than 200 €/t. WindEurope reported that mechanical recycling of wind turbine blades (WTBs) in the Netherlands costs between 150–300 €/t but the total cost ranges between 500–1,000 €/t if onsite pre-cut, transport and processing are included. Therefore, landfilling is still a viable option (WindEurope et al., 2020).

Therefore, the current legislative situation is posing a number of non-technical barriers to composite circularity which should be addressed in parallel to the technical challenges inherent to recycling. Some potential policy interventions are (EuCIA, 2022, 2024):

- establishment of dedicated waste codes to avoid heterogeneous current classification and improve quantification and traceability of waste streams;
- set of progressive targets for the reuse and recycling in addition to conceptualisation of an End of Waste regulation;
- introduction of a landfill ban to foster recycling initiatives;
- simplification of cross-boarder transport.

4 End-of-Life Management Practices in a Circular Approach

Analysing the current management of composite waste, the European Composites Industry Association (EuCIA) estimates that 40–70% of composite material waste in Europe is still disposed of in landfill or treated by incineration without energy recovery. As of 2022, the European recycling capacity for composite materials was limited to about 50 kt/year (EuCIA, 2022), compared to a generation of 800 kt/year (Assoambiente, 2023).

In addition to issues related to the legislative framework and other practical challenges such as lack of standardised procedures for the collection and separation of this type of waste in different sectors, immature market for secondary materials and limited value of recycled composited due to low-cost virgin alternatives, resulting in a limited number of dedicated recycling facilities (EuCIA, 2022, 2024), one fundamental barrier to recycling of composites lies in their nature. In fact, they are meant to have very good resistance and long lifetimes, but these characteristics are an obstacle when it comes to recycling at the end of the useful life.

Nevertheless, the increasing interest for this waste stream led to several recycling technologies for composite materials to be investigated: mechanical, thermal and chemical processes, presenting various technology readiness levels (TRL), have been tested to improve material reclamation and quality of recovered materials (Beauson et al., 2022; Fonte & Xydis, 2021). In addition, some waste prevention measures, eco-design practices and repurposing options have been considered in the context of the circular economy.

4.1 Prevention and Repurposing

Before the wind blades reach their end of life, actions can be implemented to prevent the generation of waste (i.e., reuse, reconditioning, remanufacturing, and repurpose). Another strategy to tackle with the issues of composite waste is to avoid its generation by applying alternative materials that are more recyclable, such as natural materials (wood) and thermoplastic composites (Khalid et al., 2023; Rathore & Panwar, 2023). Otherwise, following the eco-design approach, it would be suggested to design the composite materials so that the polymer matrix is easily removable, biodegradable or reusable. This can be done by introducing thermoplastics matrices, instead of thermosets, or by using recyclable thermosets, instead of common epoxy or polyesters (Mishnaevsky, 2021).

If the application for the original purpose is not possible, the subsequent action relies on reclaiming the material components of the WTB. On a large-scale view, these wastes could be repurposed and be used in the construction industry or in innovative civil engineering projects like urban furniture, bridges, children's playparks, sound barriers, bus and bicycle shelters (Khalid et al., 2023). Repurposing options only delay the problem of composite waste management, but meanwhile the value of composite material is maintained for longer and the production of raw materials needed for those applications is avoided.

4.2 Recycling

Expecting an increasing relevance of composite materials waste and especially of wind turbine blades in the coming years, several recycling technologies of composite materials that have been under development in the past decade are gaining importance. The processes can be roughly divided into mechanical, thermal and chemical (Fig. 4).

Mechanical Recycling. Mechanical recycling refers to processes which reduce the waste composite to smaller particles and is generally based on shredding and/or grinding. The outputs are resin-rich powder and fibre-rich products that can be directly reused as a filler and as reinforcement (De Fazio et al., 2023; Fonte & Xydis, 2021).

Generally, the process involves one or more size-reduction stages (cutting, shredding, crushing, and grinding), resulting in particles of reduced dimensions composed of a mixture of fibres and resin. Subsequently, the crushed material is separated in two or more fractions based on the size, by sieving, or on other characteristics, e.g., weight, by using air classifiers or zig-zag classifiers (Gonçalves et al., 2022). Typically, the finer fractions consist mainly of resin, while the coarser components consist of fibres embedded within the resin and can be mixed with other materials or impurities.

Mechanical recycling can be based on several commercially available technologies (Colledani & Turri, 2022). Shredding and hammer milling are the most diffused methods for the mechanical recycling of composite materials (De Fazio et al., 2023). With the aim of controlling the final size of the recovered particles, as well as of maximizing the throughput, the process can be optimised. For example, it can consist of two steps: a preliminary coarse shredding using a single shaft shear shredder to reach

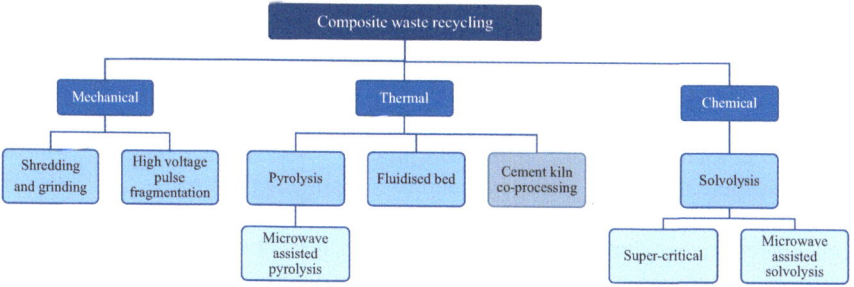

Fig. 4 Recycling methods for composite materials waste described in the following sections

particles size below 10 mm, followed by a cutting mill which is responsible for the final size distribution (Colledani & Turri, 2022).

Since it is based on composite fragmentation, the main disadvantage of mechanical recycling is that it reduces the size of the recovered fibres and therefore limits the use of these materials in applications where high performances are required. Theoretically, the shredding powder can be reused both as filler and as reinforcement; however, its application as filler is still limited both in terms of weight fraction that can be incorporated (maximum 10%) and in terms of economic convenience, since it results more expensive than virgin alternatives such as calcium carbonate or silica (Colledani & Turri, 2022). The coarser fractions containing the fibres, instead, can be reused directly as reinforcement in new composites; applications with thermoplastic polymers are more successful compared to thermoset composites as the latter show reduced bonding between recycled particles and new resins, compromising the final mechanical properties (Colledani & Turri, 2022).

All in all, secondary applications include concrete reinforcement (mainly in case of GF), replacement of virgin reinforcement fibres in new composites by shredded composites, reuse to develop products for different applications like dough moulding compound (DMC) and bulk moulding compound (BMC). The crushed and separated materials are thus used in new products such as thermal insulation panels, sound-absorbing panels, construction panels, or as filling or reinforcing materials in the production of cement or asphalt (Jani et al., 2022; Khalid et al., 2023).

Nowadays, mechanical recycling is the most common process to treat composite waste, especially for GFRP for which TRL 9 has been reached, while for CFRP the TRL is still at lower levels (6–7) (WindEurope et al., 2020). Nevertheless, mechanical recycling process is considered downcycling due to the reduced properties of the recycled material.

Mechanical processes for diminishing size can be also employed as pre-treatment before other recycling or recovery stages (Sommer et al., 2022).

In addition to conventional mechanical processes, high-voltage fragmentation (HVF) has been tested on composites as innovative process, although it was traditionally used in rock mining applications (Mativenga et al., 2016). This electro-mechanical process consists of placing the material in a dielectric ambient, such as de-ionized water, and inducing a high-intensity and fast-growing voltage (80–200 kV, at a pulse less than 500 ns). This generates a high electric field and results in dielectric breakdown creating a spark plasma channel travelling between material internal boundaries and weak regions. The spark channel generates a shockwave with high pressure (10^{10} Pa) and temperature (10^4 K) inside the material and applying several electrical pulse that the material finally disintegrates (Colledani & Turri, 2022; Mativenga et al., 2016). This process works directly on the interfaces between different materials i.e., resin and fibres, allowing the recovery of cleaner fibres with reduced resin residues. Moreover, wider fibre length distribution and higher percentage of fibres at mean fibre length have been observed. However, energy requirements are higher than conventional mechanical processes (Mativenga et al., 2016).

Thermal Recycling. Thermal recycling is a group of recycling processes involving the application of heat to separate the fibres from the matrix. These techniques allow to recover both fibres and energy, depending on the process, in terms of heat or some oil fractions from the combustion of the matrix material. The thermal degradation of the plastic matrix occurs at relatively low temperature, ranging from 450 to 800 °C depending on the type of polymer composing the matrix. This category of recycling processes includes various types of treatment such as pyrolysis, also microwave assisted, and combustion in fluidised bed.

Pyrolysis. In pyrolysis the composite materials are heated up to 450–700 °C in the absence of oxygen enabling the volatilization of the polymer matrix into gas and oil and thereby the liberation of the fibres and fillers, if present. This process converts the matrix into gases (mainly CO_2, CO, hydrogen, and light hydrocarbons), oils (a mixture of liquid organic substances), and solid products (fibres and char) (Colledani & Turri, 2022; De Fazio et al., 2023). The temperature depends on the nature of the resin: polyester resins need a lower temperature, while epoxy or high-performance thermoplastics need higher temperatures (Gonçalves et al., 2022).

After the pyrolysis vessel, a condensing chamber is placed to recover condensable gases into liquid form, while combustible gases can be fed back to the reactor. In fact, the recovered gases and oils can be used as fuel to power the pyrolysis itself, compensating the high energy demand (ETIP Wind, 2019); otherwise, the oils can be further processed in refinery to obtain polymers or monomers for producing resin again or can be used as liquid fuel for other purposes (Gonçalves et al., 2022). The solid residues, instead, consist of the fibres and char, i.e., carbonaceous residues, which are deposited on their surface and need to be purified from. Generally, a thermal post-treatment is conducted at lower temperature (450 °C) but there is a trade-off between resulting mechanical properties and the amount of remaining resin residues, since the thermal process causes further degradation of the fibres (Gonçalves et al., 2022). Pyrolysis is currently mainly applied to CFRP since it is more resistant to high temperatures. However, in order not to compromise the strength of the carbon fibres, the upper limit of temperature is between 500 and 550 °C. Also, from the economic point of view, considering high investment and running costs, pyrolysis is a viable option only for CFs (Colledani & Turri, 2022). Figure 5 illustrates the pyrolysis process.

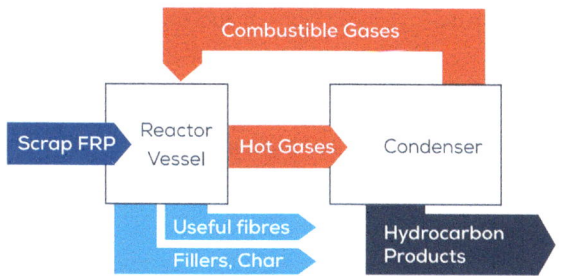

Fig. 5 Schematisation of pyrolysis process (WindEurope et al., 2020)

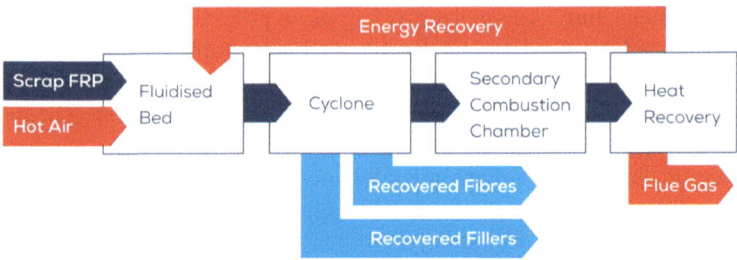

Fig. 6 Schematisation of fluidised bed process (WindEurope et al., 2020)

Combustion in fluidised bed plants. The fluidized bed combustion process implies a bed of solid particles, i.e., silica sand sized around 0.85 mm, heated and transformed into a fluid state by suspension in a stream of hot air (between 450 and 550 °C), at a fluidisation velocity between 0.4 and 1.0 m/s (Gonçalves et al., 2022). The composite material is fed in the fluidised bed allowing the rapid heating of the particles and enabling the release of fibres through friction with the sand. The fibres are subsequently separated from the gaseous flow through a subsequent gas–solid separation technology such as a cyclone (Gonçalves et al., 2022). The gaseous flow contains the organic fraction of the resin which can be sent to energy recovery in a secondary combustion chamber at a higher temperature (1000 °C) (Fonte & Xydis, 2021; Gonçalves et al., 2022). However, the friction with sand as well as the high temperature cause damages to the fibres and consequently the reduction of their tensile strength: GFs loose up to 50% of their mechanical properties (flexural or tensile strength), while CFs only 25% (Colledani & Turri, 2022; De Fazio et al., 2023; Fonte & Xydis, 2021). Moreover, the process requires shredded material as input, resulting in the recovery of short fibres only (De Fazio et al., 2023). Figures 6 and 7 report a simplified representation of the process and the necessary equipment, respectively.

The presence of the bed allows a high efficiency of heat transfer, the process allows for almost total removal of the matrix. However, fibres exhibit severely reduced mechanical properties due to various factors, including temperature (which is selected depending on the resin), oxidation, and the abrasion effect caused by the sand.

Microwave-assisted pyrolysis. The main difference between conventional and MW-assisted pyrolysis lays in the thermal transfer process: thanks to MWs the composites are heated directly in their core, instead of their surface, resulting in energy savings and time. MW-assisted pyrolysis can be applied both to CFRP and GFRP, but the particles size needs to be reduced before entering the process. Nevertheless, there is still a reduction of mechanical properties, although more limited compared to conventional pyrolysis, and it has been observed that residues of matrix can be found on the surface of the treated fibres, resulting in poorer adhesion properties.

Chemical Recycling. Chemical recycling exploits the process of solvolysis, which is a chemical process where the resin is dissolved using a solvent at specific

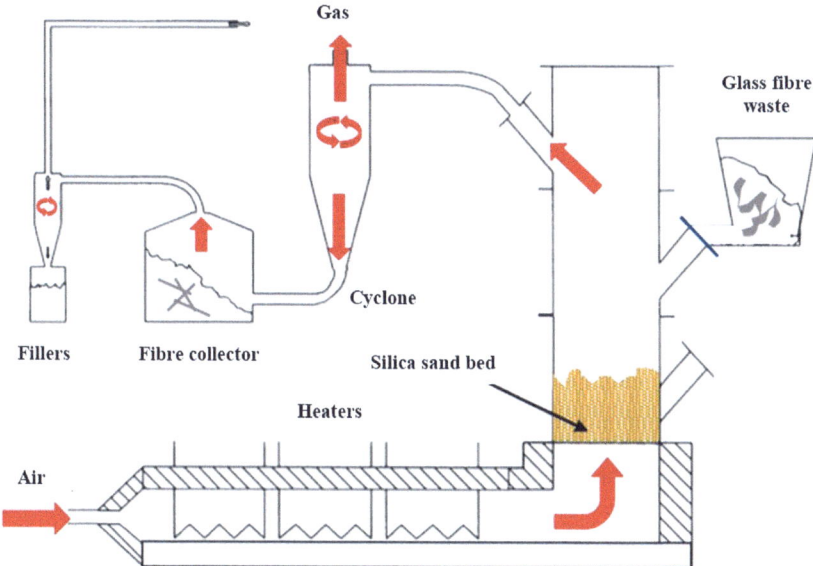

Fig. 7 Schematisation of equipment for fluidised bed recycling. Reprinted from Fonte and Xydis (2021) with permission from Elsevier

temperatures and pressure, in order to set the fibres free from the matrix bonds. In fact, the solvent can diffuse into the composite material and break the specific bonds between the matrix and the reinforcement fibres; this also allows the recovery of monomers from the polymeric matrix and to avoid the formation of char typical of thermal processes.

The main process parameters are the solvent used (preferably environmentally friendly and reusable), the process temperature, pressure, and catalytic substances. Several processes based on various solvents have been studied to cover a range of thermoplastics and thermosets composite materials; depending on the nature of the solvent, solvolysis takes on a different name: hydrolysis with the use of water, glycolysis with the use of glycols, methanolysis with the use of methanol, or acid digestion if using acids. In addition to the substances just listed, other alcohols can also be used as solvents (e.g., ethanol, propanol, and acetone). In any case, water is the most used, either alone or with a co-solvent (Colledani & Turri, 2022; Mossali et al., 2020).

To activate the process, the activation energy of the process must be reached through increased temperature or the use of catalysts, e.g., alkaline compounds such as potassium hydroxide (KOH) and sodium hydroxide (NaOH) (De Fazio et al., 2023).

Moreover, the process can be conducted in subcritical or in supercritical conditions (De Fazio):

- subcritical solvolysis is generally at atmospheric pressure and at a temperature below 200 °C; consequently, catalyst and swelling additives are required to accelerate the reaction;
- supercritical solvolysis implies high temperatures and pressures i.e. in the range of 350–450 °C and 4–27 MPa, respectively.

Supercritical solvolysis has shown better results as the working conditions improve diffusion and transport properties of the fluid, resulting in a good penetration of the solvent into the laminates and thus better removal of the matrix from the fibres by dissolution. Nevertheless, the reactors need to resist high temperature and pressure as well as corrosion which can be caused by solvents in supercritical state (De Fazio et al., 2023).

In general, sub- and supercritical solvolysis allow a very good matrix removal (over 95%) but can compromise the residual mechanical properties, such as tensile strength, of the recovered fibres (De Fazio et al., 2023). Glass fibres are more sensitive to degradation as they show fragility at high temperature and in corrosive conditions and can suffer from a severe decrease of the mechanical properties (up to 50% with respect to virgin fibres). Instead, carbon fibres maintain the original length, and the mechanical properties reduction are more limited (De Fazio et al., 2023).

In addition to the fibres, also resins and chemicals can be recovered in a liquid form: it is possible to extract plastic monomers, which are then used as fuels or as feedstock to produce new polymers (Gonçalves et al., 2022).

In contrast to supercritical solvolysis, the acid digestion can be conducted at atmospheric pressure and usually consists of a pre-treatment with acids or catalytic solutions to enhance composite swelling and therefore better penetration of the solvent. Acetic acid is the most used in a solution together with hydrogen peroxide (H_2O_2). The process can be also microwave-assisted to heat the chemical solution. Nevertheless, acid digestion requires lower processing temperature with respect to solvolysis preserving better residual mechanical properties of the fibres both for GFs and CFs while allowing a very good matrix removal (over 95%) (De Fazio et al., 2023).

The main advantage of solvolysis is that it can recover very clean fibres, i.e., almost total removal of the matrix is possible, and the damages to the fibres themselves are more limited if compared to the output of thermal and mechanical treatments, e.g., full length can be preserved. Also, resin monomers can be recovered as chemical building blocks. Moreover, the process requires lower temperature with respect to thermal recycling. Also, energy demand to assure severe processing conditions is significant (from 20 to 90 MJ/kg). Therefore, solvolysis is economically viable mainly for CFs that have higher commercial value than GFs, although in principle the process is applicable to both types of fibres (De Fazio et al., 2023).

Cement Co-processing. When applied in cement kilns, the combustion of composites combines both material and energy recovery and the process is commonly addressed as cement co-processing. In fact, mechanical grinded GFRP has been used in cement kiln to recover both the reinforcing fibres composed of inert material and the energy from the matrix (Fonte & Xydis, 2021). In practice, the organic content of the blade waste, represented by the polymeric matrix, is recovered as thermal energy, substituting partially the traditional fossil fuels (usually pet-coke), while the mineral fraction of the waste is integrated as ash in the cement, providing useful mineral raw material for clinker production (i.e., the initial stage of cement production). Nevertheless, fibres are losing their original physical shape, and this prevents further recycling.

Figure 8 depicts a simplified version of co-processing of composite waste in a cement kiln.

The advantage of utilizing the composites as filler in cement is that other raw materials, as well as fossil fuels, can be saved. In terms of raw materials, blade waste can provide CaO, SiO_2 and $Al_2O_3 + Fe_2O_2$, the main ingredients for clinker production, which are also found in the glass fibres. Table 4 compares the requirement and the average glass fibres composition.

According to Schindler et al. (2024), around 479 kg of raw materials can be displaced per t of fed WTB waste: 110 kg of CaO, 290 kg of silica, 70 kg of Al_2O_3 and 9 kg of Fe_2O_3, in addition to 490 kg of coal. This result is in line with previous data that reported a potential saving of 461 kg of raw materials when one tonne of blade waste is co-processed in cement kilns (WindEurope, 2021b).

Additionally, feeding calcium oxide directly with raw materials allows to avoid its production from $CaCO_3$ by the calcination reaction and therefore it is possible to reduce both CO_2 produced by the reaction itself and CO_2 related to fuel combustion. Schindler et al. (2024) reported that for each kilogram of CaO delivered by raw material already in this form, 1.24 kg CO_2 can be saved. They also reported that, based on the composition found in the literature, WTB glass fibres contain on average 20% of CaO.

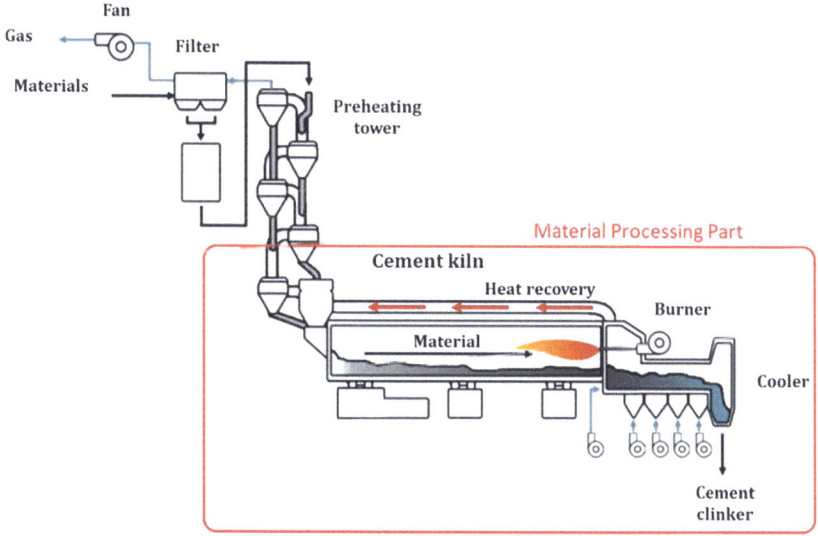

Fig. 8 Representation of composite waste co-processing in a cement kiln. Reprinted from Fonte and Xydis (2021) with permission from Elsevier

Table 4 Comparison between clinker requirements and average composition of glass fibres and E-glass

	CaO (%)	SiO$_2$ (%)	Al$_2$O$_3$ (%)	Fe$_2$O$_3$	B$_2$O$_3$	MgO	Sources
Clinker requirements	65–75	21	6–11		x		Nagle et al. (2020)
Average composition of WTB glass fibres	20	58	13	0.5%	6.5%	4%	Schindler et al. (2024)
Average composition of E-glass	32–38	52–58	5–15				Nagle et al. (2020)

Note E-glass is electrical glass which is typically used in electrical components and reinforced plastics (Schindler et al., 2024)

So, it is self-evident that WTB co-processing leads to a reduction of the carbon footprint of cement; however, its quantification is not straightforward, since different information can be found:

- according to Liu et al. (2019), each tonne of blade waste can replace 600 kg of coal fuel, equivalent to 4.16 GJ of energy;
- WindEurope et al. (2020) reported WTBs co-processing can reduce CO_2 emissions by 110 kg;
- it was reported that WTB co-processing is reducing carbon footprint of cement production by 16% (Colledani & Turri, 2022; WindEurope et al., 2020);
- based on the data reported by Schindler et al. (2024), assuming that CaO content in GFRP waste is 20%, feeding one tonne of end-of-life (EoL) WTBs could save around 248 kg CO_2.

However, it has been observed that a maximum of 10% of the fuel can be substituted by GFRP since the presence of boron in GFs, generally made of E-glass, compromises the performance of cement (Pickering, 2006). Moreover, the material needs to be mechanically pre-treated to reduce its size and cannot contain impurities (Gonçalves et al., 2022). In fact, some other components present in EoL wind turbine blades need to be removed in the pre-processing to avoid adverse effects on cement quality and issues to the processing. Especially PVC needs to be removed due to its chlorine content, that needs to be limited to 0.2%, as well as metals parts and copper, sometimes present as lightning protection (Schindler et al., 2024).

Considering that cement co-processing allows the combination of energy recovery from the resin and material recovery from the fibres which become part of the cement itself, this treatment route is receiving special attention by industry associations. Co-processing in cement kiln seems to be an interesting solution in the short and medium term, with the major aim of avoiding landfilling, as it is already commercially available to process large volumes of composite waste. For example, a cement kiln of Holcim Group in Northern Germany with a total co-incineration capacity of 30,000 t/year is already treating about 10,000 t/year of waste coming from wind turbine blades with a gate fee of around 150 €/t (WindEurope, 2021b; WindEurope et al., 2020). However, the industry associations support also the development of alternative recycling technologies that can provide recyclates retaining higher value and applicable in new composite manufacturing.

4.3 Performance, Benefits, Drawbacks

In Table 5 the main strengths and drawbacks of the presented technologies are summarised.

Review of Life Cycle Assessment on EoL Management of Composite Waste. With the aim of comparing the presented recycling and recovery technologies from the environmental point of view, a literature review has been made looking for life cycle assessment studies. It can be noted that at the moment only few LCA studies focusing (or including) the end-of-life stage of wind turbine blades have been published and most of them are focused only on energy requirements and/or carbon footprint. The following paragraph summarises the fundamental information from a selection of LCA studies on EoL wind turbine blades focusing on mechanical, chemical and thermal recycling techniques.

The reviewed studies investigate the environmental impact of various EoL treatments of WTBs, focusing on mechanical, chemical, and thermal recycling techniques.

Liu et al. (2019) assess nine EoL processes for WTBs, including landfill, incineration, mechanical recycling, fluidised bed pyrolysis, microwave-assisted pyrolysis, chemical recycling (hydrolysis and solvolysis), high-voltage fragmentation (HVF), and blade life extension, applied to three blade types: glass fibre (GF), carbon fibre (CF), and hybrid blades. Pyrolysis offers full resin recovery but yields the highest environmental burden, while solvolysis and mechanical recycling emerge as the best options for hybrid and CF blades due to their net environmental benefits. The study finds that fluidised bed and conventional pyrolysis have the highest environmental impacts, followed by chemical recycling and high-voltage fragmentation (HVF). Interestingly, landfill and incineration show lower impacts than expected. For GF blades, solvolysis and mechanical recycling are the most beneficial, whereas fluidised bed pyrolysis results in negative net impacts. For hybrid GF/CF blades, chemical recycling and HVF are more environmentally favourable, while CF blades show significant benefits across all recycling options, particularly with solvolysis, which yields the highest environmental gains.

Nagle et al. (2020) compare the environmental performance of three scenarios for blade waste management in Ireland: landfill, co-processing in German cement plants, and co-processing in hypothetical Irish cement plants. Co-processing significantly reduces damage, particularly by replacing coal and raw materials. The results reveal that the landfill scenario had the worst performance across all environmental categories, mainly due to the lack of material recovery benefits, and that the

Table 5 Summary of possible recycling methods of composite waste and the relative advantages and disadvantages; authors' elaboration based on (Gonçalves et al., 2022; Khalid et al., 2023; WindEurope et al., 2020)

Method	Advantages	Disadvantages
Mechanical recycling (shredding, grinding)	High treatment capacity Low energy consumption Low costs Industrial scalability Possible on-site processing TRL 9 for GFRP	Not complete separation between matrix and fibres Recovery of short fibres: the process disrupts their physical integrity Low quality output materials: reduction of mechanical characteristics such as strength, reduction in chemical-physical properties, possible content of other materials and impurities Low commercial value of the recycled material, not competitive to virgin raw materials Generation of process waste (up to 40%) Generation of dust during the operations requires conducting the processes in a closed and controlled environment (Rathore & Panwar, 2023) TRL 6–7 for CFRP
High voltage fragmentation	Cleaner fibres (lower amount of residual resin) compared to conventional mechanical recycling Wide range of fibres length Scalability potential Low investment to reach next TRL	High cost Low throughput Decreased quality of the recovered fibres (fragmented) TRL 6 Only laboratory and pilot-scale equipment Higher energy consumption compared to conventional mechanical recycling

(continued)

Table 5 (continued)

Method	Advantages	Disadvantages
Pyrolysis (also MW-assisted)	Recovery of long fibres Byproducts (syngas and oils) can be used as energy source or as new polymers building blocks; potential internal energy recovery to self-sustain the process No need for chemicals TRL 9 Technology easily scalable MW-assisted shows lower damages to fibres and is easier to control	Reduction of mechanical properties of fibres (e.g., tensile strength, Young's modulus, flexural strength) Char residues on fibres requires thermal post-treatment causing further strength loss Limited secondary applications of fibres Energy-intensive Long processing times High investment and running costs Economically viable only for CFs due to higher commercial value than GFs, which is also more sensitive to high temperatures MW-assisted pyrolysis: TRL 4–5 Presence of matrix residues on fibres in case of MW-assisted process
Combustion in fluidised bed	Efficient heat transfer Almost total removal of matrix from fibres, without char residues Possible energy recovery by post-combustion of gases Tolerant to contaminated or mixed composite waste without need for pre-processing	Recovery of short fibres Severe reduction of mechanical properties, especially due to friction with sand More suitable for CFs than GFs Requires mechanically pre-treated input Still needs to be scaled-up TRL 5–6 Flue gas to be treated
Solvolysis	Recovery of clean and long fibres (almost total removal of the matrix is possible and full length can be preserved) Lower damages to fibres compared to thermal treatments Lower temperature with respect to thermal treatments Possible use of low-risk solvents such as water Recovery of resins monomers as chemical building blocks	Requires expensive reactors (stainless-steel pressure vessels) Significant energy demand Higher process costs than thermal treatments Use of large volumes of chemicals Would be economically viable for CFs due to higher commercial value than GFs Potential health and environmental risks due to solvents and catalysts Still studied at lab-scale TRL 5–6

(continued)

Table 5 (continued)

Method	Advantages	Disadvantages
Supercritical solvolysis	Better quality of recovered fibres compared to conventional solvolysis Less toxic process	Requires reactors resistant to corrosion
Cement co-processing	High treatment capacity Efficient and fast treatment Already at industrial scale TRL 9 Potential reduction of carbon footprint and raw material requirements for cement production Avoids solid residues production	Requires mechanically pre-treated input Requires removal of foreign materials potentially bound to composite waste Not proper recycling of fibres due to loss of original physical shape Only suitable for GFs Pollutants and particulate matter emissions to be treated

TRL = Technology Readiness Level; GFRP = Glass Fibre Reinforced Plastic; CFRP = Carbon Fibre Reinforced Plastic; MW = Microwave; GF = Glass Fibre; CF = Carbon Fibre

local treatment scenario performs better than exporting waste due to transportation impacts.

Sommer et al. (2022) evaluate the influence of political regulations on waste management for glass fibre-reinforced polymer (GFRP) and carbon fibre-reinforced polymer (CFRP) in Europe.

The life cycle assessment included several EoL options such as mechanical recycling, solvolysis, HVF, and pyrolysis, with further treatments like blending or briquetting before co-processing in cement kilns or incineration. The analysis shows that HVF had the highest environmental impacts, followed by solvolysis, while thermal treatments show moderate impacts partially offset by energy recovery. Mechanical recycling shows significantly lower impacts. The study highlights the importance of recycling credits from secondary applications, especially from avoiding the production of virgin CFs, which yields the highest benefits in terms of reduced greenhouse gas (GHG) emissions. For CFRP waste, chemical recycling is the preferred treatment method, while for GFRP, mechanical recycling performs best, particularly in generating high-quality epoxy resin from recovered materials.

Diez-Canamero and Mendoza (2023) analyse the circular economy performance and carbon footprint of seven EoL options, including landfill, incineration, repurposing, solvolysis, pyrolysis, grinding, and co-processing. They find that repurposing and solvolysis are the most environmentally beneficial thanks to credits associated with high-quality material recovery. Pyrolysis, on the other hand, has the worst carbon footprint due to its energy-intensive nature, despite offering energy recovery. Mechanical grinding and co-processing, less impactful, perform better than landfill and incineration, which provides no significant environmental benefits.

Gennitsaris et al. (2023) investigate eleven EoL scenarios for wind turbines in Greece. The study compares various processes such as repurposing, mechanical recycling, and both conventional and microwave-assisted pyrolysis. Conventional pyrolysis shows high GHG impacts due to its energy demands, while microwave-assisted pyrolysis reduces land occupation and energy consumption. The study concludes that while microwave-assisted pyrolysis has lower environmental impacts compared to conventional methods, repurposing offers an efficient alternative with significant environmental savings.

Across the studies, mechanical and chemical recycling, particularly solvolysis, consistently show the best environmental outcomes due to high material recovery, while pyrolysis, especially in conventional form, is energy-intensive and less beneficial in terms of net environmental impact.

Figure 9 reports the energy consumption in the production of the virgin raw materials for composites i.e., carbon and glass fibres as well as epoxy and polyester resin, in order to be compared to the range of energy requirements reported for different composite waste recycling technologies.

The amount of energy needed for the primary production of reinforcing fibres (especially carbon fibres) and resins is generally higher than the energy demands of recycling processes. Mechanical recycling shows the lowest energy demand, ranging between 0.17 and 4.8 MJ/kg of processed waste, while other treatments yield more variable results depending on the source of the data. The highest energy consumption is reported for chemical recycling at 91 MJ/kg.

Moreover, Fig. 10 depicts the comparison between process costs and potential secondary material value for different recycling options.

When treating carbon fibres, the potential revenue from the recovered material appears to outweigh the recycling costs, particularly in the case of solvolysis and pyrolysis. Additionally, mechanical grinding and co-processing of glass fibre reinforced plastics show potential for economic sustainability. Pyrolysis is the only method that presents data for both fibre types and seems to be a viable option primarily for carbon fibres.

5 Industrial Plants and Initiatives

There is no doubt that EoL WTB recycling and enhanced material recovery has been gaining importance. Organisations such as the EIT RawMaterials and the European Union through the Horizon Europe initiative have been fostering and funding several research projects on the subject. A non-exhaustive list of past and present projects on the topic can be found in "Accelerating Wind Turbine Blade Circularity" (WindEurope et al., 2020) and in "Polymer Composites Circularity" (SusChem, n.d.). On the other hand, the number of plants operating recycling processes of composite materials at the industrial scale is still limited. The European Composites Industry Association (EuCIA) has recently started to gather data and made available the European Composites Recycling Solutions Database.

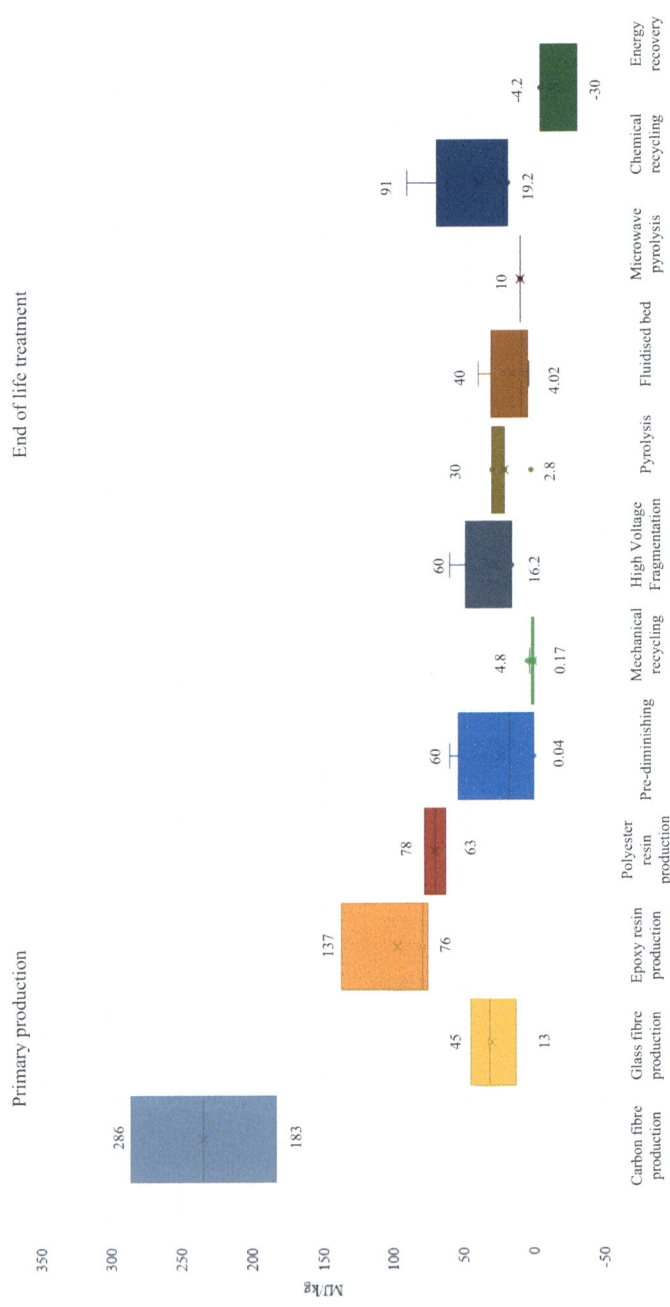

Fig. 9 Energy requirements for manufacturing of the main blade materials (reinforcing fibres and resins) and for recycling processes. Compiled by the authors based on Fonte and Xydis (2021), Jani et al. (2022), Rathore and Panwar (2023)

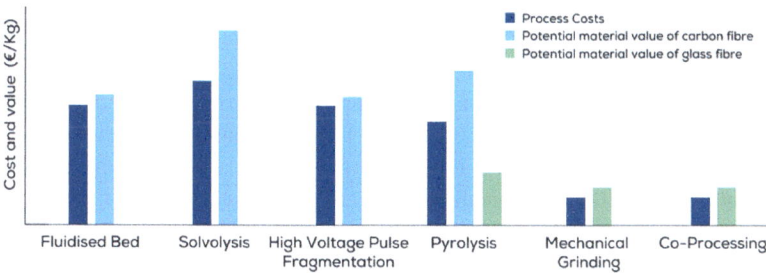

Fig. 10 Estimated cost and revenues associated to different composites recycling technologies (WindEurope et al., 2020)

The recycling plants across Europe vary significantly in terms of geographic distribution and the types of materials they process. Italy, Germany and the UK report the highest number per country in Europe (4–5 each). France, Denmark and Switzerland also host some plants (2–3 each), while for other countries such as Austria, Belgium, Spain and Finland only one recycling plant is reported.

These plants handle various types of waste, including thermoset composites, fibrous waste (glass or carbon fibres), and composite waste from industrial production.

The recycling processes are various, with mechanical recycling being the most widespread due to its cost-effectiveness and scalability (17 out of 26 companies), followed by thermal methods provided by 10 companies. Chemical recycling, which allows for the high-purity recovery of fibres, is gaining attention but so far only four of the reported companies provide chemical solutions.

The types of fibres processed in these facilities also differ, with carbon fibre being more commonly recovered than glass fibre. Some plants are also able to process both types of reinforcing fibres.

Examples of facilities providing mechanical recycling of thermoset composites (both GFRP and CFRP) are Roth International in Germany and Gees in Italy.

Kuusakoski Oy in Finland treats composite materials by shredding and then supplies it to the cement industry for co-processing.

Composite Recycling in Switzerland relies on pyrolysis for processing only industrial production waste, as its composition is well known, while for wind turbine blades the presence of PVC is of concern. The primary outputs are secondary glass fibres and carbon fibres to a lower extent, but also oil for petrochemical industry.

Reprocover in Belgium and Continuum in Denmark provide mechanical processing of composite waste from wind turbines and produce secondary materials for the construction sector. Anmet in Poland provides thermal recycling of wind turbine blades addressing the recovery of carbon fibres.

More industrial and pilot initiatives are being developed in relation to research projects constantly striving for technological advancements in the field.

6 Conclusions

Wind turbine blades are generally made of composite materials designed to withstand time and harsh conditions but not optimized for recyclability. While up to 85% of a wind turbine is recyclable, fibre-reinforced plastic used mainly in the blades is a significant challenge both technically and from the legislative point of view.

Currently, a specific code in the European List of Waste is missing and prevents the precise quantification of composite waste streams. Also, many European countries lack specific guidelines and legislation, although harmonization could improve end-of-life management.

So far, recycling technologies for composite materials are still under development and can be divided into three categories: mechanical, thermal and chemical. Mechanical treatments, already at TRL 9, produce a resin-rich powder and a fibre-rich coarse fraction. This method is the most common but limits secondary applications due to reduced fibre size. Thermal and chemical recycling aim to release reinforcing fibres by thermally degrading or dissolving the plastic matrix These methods can also recover the polymeric matrix, which may be used for energy production or as feedstock for new polymers. Compared to mechanical methods, thermal and chemical recycling can nearly completely remove the matrix and recover longer fibres. However, these fibres often have reduced mechanical properties due to high temperatures or sand friction in fluidized beds. Pyrolysis, at TRL 9, is economically viable mainly for carbon fibres, which have a higher commercial value than glass fibres, which are more sensitive to degradation. Solvolysis is still at the lab scale (TRL 5–6) and is currently too expensive for widespread use, being more suitable for carbon than for glass fibres. Moreover, cement co-processing is a promising solution at the industrial scale, especially for the short-medium term. It combines both material and energy recovery by using the polymeric matrix as thermal energy and integrating the mineral fraction into the clinker, reducing fossil fuel usage and raw material needs. The number of operating plants for recycling of composite materials at industrial scale is still very limited. However, various industry associations related to the world of composite materials are showing great interest in developing a recycling market for these materials.

Competing Interests The authors have no conflicts of interest to declare that are relevant to the content of this chapter.

References

Assoambiente. (2023). L'Italia che Ricicla 2023. https://assoambiente.org/files/rapporto-italia-che-ricicla-2023.pdf.
AVK. (2024). The European market for fiber-reinforced plastics/composites 2023. www.avk-tv.de.
Beauson, J., Madsen, B., Toncelli, C., Brøndsted, P., & Ilsted Bech, J. (2016). Recycling of shredded composites from wind turbine blades in new thermoset polymer composites. *Composites Part A:*

Applied Science and Manufacturing, 90, 390–399. https://doi.org/10.1016/J.COMPOSITESA. 2016.07.009.

Beauson, J., Laurent, A., Rudolph, D. P., & Pagh Jensen, J. (2022). The complex end-of-life of wind turbine blades: A review of the European context. In *Renewable and sustainable energy reviews* (Vol. 155). https://doi.org/10.1016/j.rser.2021.111847.

Carrara, S., Alves Dias, P., Piazzotta, B., & Pavel, C. (2020). Raw materials demand for wind and solar PV technologies in the transition towards a decarbonised energy system. https://data.eur opa.eu, https://doi.org/10.2760/160859.

Carrara, S., Bobba, S., Blagoeva, D., Dias, A., Cavalli, P., Georgitzikis, A., et al. (2023). *Supply chain analysis and material demand forecast in strategic technologies and sectors in the EU-a foresight study.* JRC Science for Policy Report. EU CRM. https://doi.org/10.2760/334074.

Colledani, M., & Turri, S. (2022). Systemic circular economy solutions for fiber reinforced composites. In M. Colledani, & S. Turri (Eds.), Springer International Publishing. https://doi.org/10. 1007/978-3-031-22352-5.

Compositi Magazine. (2023). EMPHASIZING, un progetto per il riciclo dei compositi in fibra di vetro. https://www.compositimagazine.it/emphasizing-un-progetto-per-il-riciclo-dei-compositi-in-fibra-di-vetro/.

Cooperman, A., Eberle, A., & Lantz, E. (2021). Wind turbine blade material in the United States: Quantities, costs, and end-of-life options. *Resources, Conservation and Recycling, 168*(2022), 105439. https://doi.org/10.1016/j.resconrec.2021.105439.

De Fazio, D., Boccarusso, L., Formisano, A., Viscusi, A., & Durante, M. (2023). A review on the recycling technologies of fibre-reinforced plastic (FRP) materials used in industrial fields. *Journal of Marine Science and Engineering, 11*(4), 851. https://doi.org/10.3390/jmse11040851.

DecomBlades. (n.d.). DecomBlades. Retrieved June 6, 2024, from https://decomblades.dk/.

Diez-Cañamero, B., & Mendoza, J. M. F. (2023). Circular economy performance and carbon footprint of wind turbine blade waste management alternatives. *Waste Management, 164,* 94–105. https://doi.org/10.1016/J.WASMAN.2023.03.041.

ETIP Wind. (2019). An overview of composite recycling in the wind energy industry. https://ec.eur opa.eu/research/participants/documents/downloadPublic?documentIds=080166e5c7c5bfb8& appId=PPGMS.

EuCIA. (2022). EuCIA: Background document on circular economy. https://glassfibreeurope. eu/wp-content/uploads/2022/11/221031_EuCIA_Background_Document_Composites_Circul arity_Final_002_.pdf.

EuCIA. (2024). Enabling circular composites starts with waste codes. https://eucia.eu/enabling-cir cular-composites-starts-with-waste-codes/.

Fioretti, C., Pasquale, G., Baglivo, L., Boughdachi, A., Delfrate, D., Diani, M., et al. (2023). Deliverable D1.1 of the DeremCo Project-Map of the European composite post-use streams: Waste composition/location. http://deremco.afil.it/wp-content/uploads/2023/08/DeremCo_D1. 1-2.pdf.

Fonte, R., & Xydis, G. (2021). Wind turbine blade recycling: An evaluation of the European market potential for recycled composite materials. *Journal of Environmental Management, 287,* 112269. https://doi.org/10.1016/J.JENVMAN.2021.112269.

Gennitsaris, S., Sagani, A., Sofianopoulou, S., & Dedoussis, V. (2023). Integrated LCA and DEA approach for circular economy-driven performance evaluation of wind turbine End-of-Life treatment options. *Applied Energy, 339,* 120951. https://doi.org/10.1016/J.APENERGY.2023. 120951.

Gonçalves, R. M., Martinho, A., & Oliveira, J. P. (2022). Recycling of reinforced glass fibers waste: Current status. *Materials, 15*(4), 1596. https://doi.org/10.3390/MA15041596.

IRENA. (2019). Future of wind. https://www.irena.org/publications/2019/Oct/Future-of-wind.

Jani, H. K., Singh Kachhwaha, S., Nagababu, G., & Das, A. (2022). A brief review on recycling and reuse of wind turbine blade materials. *Materials Today: Proceedings, 62*(P13), 7124–7130. https://doi.org/10.1016/J.MATPR.2022.02.049.

Khalid, M. Y., Arif, Z. U., Hossain, M., & Umer, R. (2023). Recycling of wind turbine blades through modern recycling technologies: A road to zero waste. *Renewable Energy Focus, 44*, 373–389. https://doi.org/10.1016/J.REF.2023.02.001.
Korec. (n.d.). Korec. Retrieved July 9, 2024, from https://www.ko-rec.com/.
Lichtenegger, G., Rentizelas, A. A., Trivyza, N., & Siegl, S. (2020). Offshore and onshore wind turbine blade waste material forecast at a regional level in Europe until 2050. *Waste Management, 106*, 120–131. https://doi.org/10.1016/J.WASMAN.2020.03.018.
Liu, P., & Barlow, C. Y. (2017). Wind turbine blade waste in 2050. *Waste Management, 62*, 229–240. https://doi.org/10.1016/J.WASMAN.2017.02.007.
Liu, P., Meng, F., & Barlow, C. Y. (2019). Wind turbine blade end-of-life options: An eco-audit comparison. *Journal of Cleaner Production, 212*, 1268–1281. https://doi.org/10.1016/J.JCLEPRO.2018.12.043.
Mativenga, P. T., Shuaib, N. A., Howarth, J., Pestalozzi, F., & Woidasky, J. (2016). High voltage fragmentation and mechanical recycling of glass fibre thermoset composite. *CIRP Annals, 65*(1), 45–48. https://doi.org/10.1016/J.CIRP.2016.04.107.
Mishnaevsky, L. (2021). Sustainable End-of-Life management of wind turbine blades: Overview of current and coming solutions. *Materials, 14*(5), 1124. https://doi.org/10.3390/MA14051124.
Mishnaevsky, L., Branner, K., Petersen, H. N., Beauson, J., McGugan, M., & Sørensen, B. F. (2017). Materials for wind turbine blades: An overview. *Materials, 10*(11), 1285. https://doi.org/10.3390/MA10111285.
Mossali, E., Picone, N., Gentilini, L., Rodrìguez, O., Pérez, J. M., & Colledani, M. (2020). Lithium-ion batteries towards circular economy: A literature review of opportunities and issues of recycling treatments. *Journal of Environmental Management, 264*, 110500. https://doi.org/10.1016/j.jenvman.2020.110500.
Nagle, A. J., Delaney, E. L., Bank, L. C., & Leahy, P. G. (2020). A comparative life cycle assessment between landfilling and co-processing of waste from decommissioned Irish wind turbine blades. *Journal of Cleaner Production, 277*, 123321. https://doi.org/10.1016/j.jclepro.2020.123321.
ORE Catapult. (2016). Sustainable decommissioning: Wind turbine blade recycling. https://ore.catapult.org.uk/resource-hub/analysis-reports/energy-transition-alliance-blade-recycling-report-full-version.
Pickering, S. J. (2006). Recycling technologies for thermoset composite materials—current status. *Composites Part a: Applied Science and Manufacturing, 37*(8), 1206–1215. https://doi.org/10.1016/J.COMPOSITESA.2005.05.030.
Rathore, N., & Panwar, N. L. (2023). Environmental impact and waste recycling technologies for modern wind turbines: An overview. *Waste Management and Research, 41*(4), 744–759. https://doi.org/10.1177/0734242X221135527/ASSET/IMAGES/LARGE/10.1177_0734242X221135527-FIG5.JPEG.
Schindler, A. K., Duke, S. R., & Galloway, W. B. (2024). Co-processing of end-of-life wind turbine blades in portland cement production. *Waste Management, 182*, 207–214. https://doi.org/10.1016/J.WASMAN.2024.04.033.
Sommer, V., Stockschläder, J., & Walther, G. (2020). Estimation of glass and carbon fiber reinforced plastic waste from end-of-life rotor blades of wind power plants within the European Union. *Waste Management, 115*, 83–94. https://doi.org/10.1016/J.WASMAN.2020.06.043.
Sommer, V., Becker, T., & Walther, G. (2022). Steering sustainable End-of-Life treatment of glass and carbon fiber reinforced plastics waste from rotor blades of wind power plants. *Resources, Conservation and Recycling, 181*, 106077. https://doi.org/10.1016/J.RESCONREC.2021.106077.
SusChem. (n.d.). Publications. Retrieved September 2, 2024, from http://suschem.org/publications/.
WindEurope, Cefic, & EuCIA. (2020). Accelerating wind turbine blade circularity. https://windeurope.org/wp-content/uploads/files/about-wind/reports/WindEurope-Accelerating-wind-turbine-blade-circularity.pdf.

WindEurope. (2021a). Wind industry calls for Europe-wide ban on landfilling turbine blades. https://windeurope.org/newsroom/press-releases/wind-industry-calls-for-europe-wide-ban-on-landfilling-turbine-blades/.

WindEurope. (2021b). WindEurope CEO visits German cement plant that's running on blade waste. https://windeurope.org/newsroom/news/windeurope-ceo-visits-german-cement-plant-thats-running-on-blade-waste/.

WindEurope. (2024). Wind energy in Europe: 2023 Statistics and the outlook for 2024–2030. https://windeurope.org/intelligence-platform/product/wind-energy-in-europe-2023-statistics-and-the-outlook-for-2024-2030/.

Photovoltaic Panels

Gaia Brussa

Abstract Solar energy, harnessed through photovoltaic (PV) panels, is essential in the shift toward a decarbonised energy system. The worldwide rapid growth of PV installations, with an estimated addition of 600 GW in year 2024, highlights the pivotal role of this type of renewable energy. However, since many PV modules are starting to approach the end of their 30-year lifespan, the need for effective end-of-life (EoL) management becomes critical. This chapter explores the structure and market share of key PV technologies, such as crystalline silicon (c-Si) and thin-film types like CdTe and CIGS, focusing on specific raw material demand. The importance of recycling to recover critical materials is emphasised, making a distinction between processing by laminated glass recyclers, which primarily recovers bulk materials like glass and the aluminium frame, and advanced recycling techniques that aim for higher-value material recovery. Advanced recycling is structured in three key steps: disassembly, delamination, and metal recovery. Disassembly involves the removal of frames, junction boxes, and cables. Delamination, the most challenging phase, aims at the removal of the encapsulation layer from the solar cells and exploits different processes like mechanical shredding, combustion or pyrolysis, or chemical dissolution. The final recovery step focuses on extracting valuable materials such as silicon, silver, and rare metals through processes like etching, leaching, and hydrometallurgical extraction. The chapter also outlines the legislative frameworks, particularly the EU's WEEE Directive, which governs PV waste management, and the existing recycling infrastructure in Europe.

Keywords Energy Transition · Waste · Renewable Energy · Recycling · Photovoltaic Panels

---

G. Brussa (✉)
Department of Civil and Environmental Engineering, Politecnico di Milano, Milano, Italy
e-mail: gaia.brussa@polimi.it

MatER Study Center, Laboratory for Energy and the Environment Piacenza, Piacenza, Italy

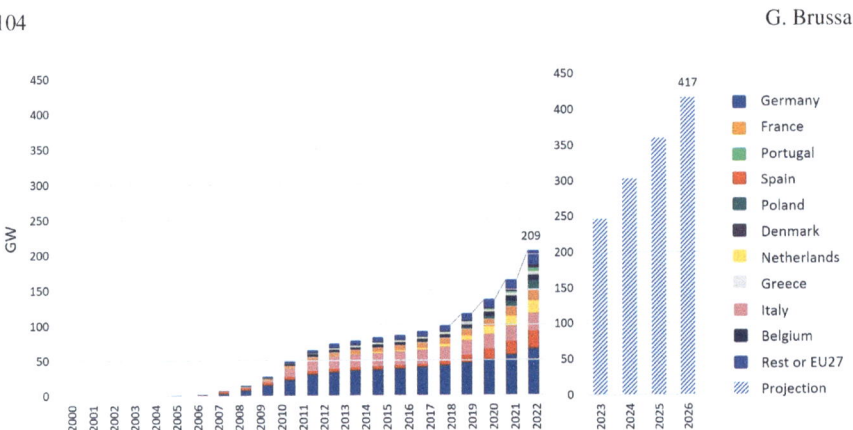

Fig. 1 Historical data and low scenario outlook of European cumulative solar PV capacity (GW). Author' elaboration based on SolarPower Europe (2022)

1 Solar Energy

Solar energy can be exploited by photovoltaic (PV) panels to produce electricity. In the last decades, solar PV installed capacity has experienced a significant increase and is expected to be one of the essential technologies for a decarbonised energy system. As depicted in Fig. 1 the PV cumulative installed capacity has surpassed 200 GW in Europe and is expected to at least double in the next years (IRENA & IEA-PVPS, 2016; SolarPower Europe, 2022). Therefore, the number of PV modules reaching the end of useful life is expected to grow.

PV modules have a useful lifespan of approximately 30 years (IRENA & IEA-PVPS, 2016), which means that EoL PV panels are expected to be an issue already in the coming years, since in most countries the first installations began in the early 2000s and continued to grow. It is expected that the number of end-of-life photovoltaic panels will soon start to increase and will continue to do so at nearly the same pace as the installations that took place in the past. These projections have driven several efforts in the research and development of advanced recycling solutions with the aim of avoid their disposal at the end of life and keep the valuable materials in the economy.

2 Characterisation of the Waste Flow

The main basic unit of a PV technology is the photovoltaic panel constituted by individual PV cells, where the conversion of solar radiation into electric energy is performed. Cells are composed by wafers, i.e., thin slices, or thin films of semiconducting materials. Single cells are then connected to form the so-called PV module. A solar PV plant requires other elements in addition to PV modules: inverters, mounting

Photovoltaic Panels

structures, and general electrical components, as well as trackers in case of solar tracking facilities.

Solar PV panels are currently based on different technologies, the most common of which are (Carrara et al., 2020):

- wafer-based crystalline silicon (c-Si), either mono-crystalline or multi-crystalline, also known as first generation PV panels (Jean et al., 2015);
- thin-film technologies that can be further distinguished in cadmium telluride (CdTe), copper indium gallium diselenide (CIGS) and amorphous silicon (a-Si). They are the second generation PV panels (Jean et al., 2015), driven by the intention to reduce costs by decreasing the materials usage (Latunussa et al., 2016a, 2016b).

Additionally, the more recent third generation PV includes, among others, dye-sensitized, organic and heavy metals (e.g., perovskite which is based on lead) photovoltaic cells (Jean et al., 2015).

2.1 Crystalline Silicon Cells

The crystalline silicon cells are composed of wafers, i.e., slices of solar-grade silicon, which is typically refined to a purity of 99.9999% (Jean et al., 2015). Silicon solar cells can be further distinguished between single- or multi-crystalline depending on the number of silicon crystals involved. Single crystal silicon can be grown starting by pure silicon through the Czochralski method or float-zone technique. Then the resulting ingots, made of single crystals of semiconductor, are sliced into wafers 150 and 180 μm thick (Jean et al., 2015). Instead, multi-crystalline cells consist of multiple fragments of silicon, also called grains. Multi-crystalline blocks are formed by casting and present crystal grains randomly oriented. Grain boundaries are defects which reduce the performance of these kind of PV module in terms of electric efficiency (Jean et al., 2015). On the other hand, single-crystalline cells are more efficient and durable, but their manufacturing processes are complicated and expensive; also, the resulting cells are very fragile (Preet & Smith, 2024). The main drawback of c-Si is the relatively poor light absorption that results in the use of thick and very pure wafers (Jean et al., 2015).

As regards the structure, in addition to the semiconductor layer which constitutes the active part in light absorption, an anti-reflective coating is usually applied on silicon wafer to enhance light absorption efficiency (Wang et al., 2024b). In order to allow the formation of an electric field and let the generated current flow to the external circuits, the front and the back of cells are connected by grid-patterns made of silver and aluminium or copper paste (IRENA & IEA-PVPS, 2016). Moreover, ribbons typically made of copper and tin–lead solder connect multiple solar cells in a PV module. The semiconductor and the electric connectors are placed between two polymeric layers, commonly made of ethyl-vinyl-acetate (EVA). The thin layers of EVA act as encapsulant since air is removed from the space within and EVA is

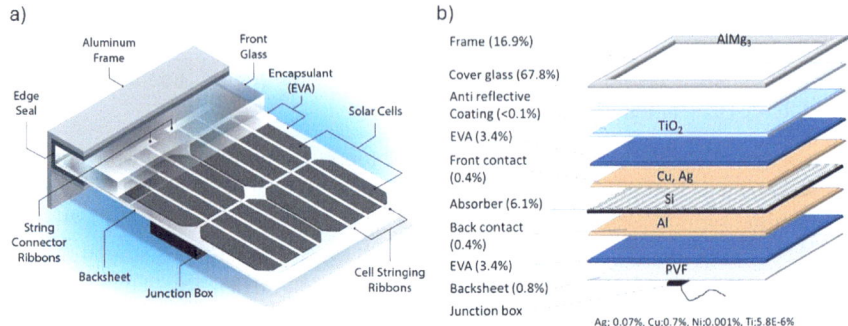

Fig. 2 Crystalline silicon (c-Si) photovoltaic cell structure, including typical composition by weight. Reprinted from **a** IRENA and IEA-PVPS (2016) and **b** Maani et al. (2020) with Elsevier permission

heated to reach the melting point and seal the structure. The front layer of the PV modules is usually made of tempered glass which is transparent to light but gives mechanical support and protects the fragile active part from adverse weather conditions such as hail. On the back of the cell there is the so-called backsheet i.e., another polymeric layer that provides electrical insulation and increases the impermeability of internal components. Generally, the backsheet consists of a fluorinated compound e.g., polyvinyl fluoride (PVF also known with the commercial name Tedlar®) and can also be made of a sandwich structure with three layers: Tedlar®/PET/Tedlar®. By connecting glass, cells and backsheet together, EVA enhances resistance to moisture, insulation, and mechanical strength of the solar cell. This laminated assembly is finally bundled together by an aluminium frame (Wang et al., 2024b).

The typical structure of a c-Si PV module is shown in Fig. 2.

2.2 Thin-Film Cells

Thin-film PV panels are a more complex technology but have the key advantage of using less semiconducting material, thanks to the fact that it is deposited in thin layers onto another substrate e.g., glass, polymers, or metals. In the commercial thin-film PV panels the active materials are more efficient in absorbing light, 10–100 times more than silicon, and therefore the absorption layer can be just a few μm thick. The most common active materials are amorphous silicon (a-Si), cadmium telluride (CdTe) or compounds of indium, selenide and gallium. In addition to extreme thinness of absorption layers, these technologies result in being lighter and cheaper, they require less energy in the production stage and rely on less expensive materials (Jean et al., 2015).

Amorphous Silicon (a-Si). Amorphous silicon cells use a non-crystalline form of silicon as active material for light absorption and energy conversion. Conversely to crystalline silicon, a-Si does not present the long-range order in the atomic structure but only on a short range. This technology was firstly introduced as an alternative to the crystalline silicon and in the early 2000s it represented the most widespread in the thin-film PV market; nevertheless, due to its low efficiency, it lost importance (Carrara et al., 2020). Although it could be used in some small-scale and low-power applications (Jean et al., 2015), this technology is considered to be phased out and therefore neglected in future projections of the PV market (Carrara et al., 2023).

Cadmium Telluride (CdTe). Cadmium Telluride PV panels are considered a thin-film technology because the thickness of the active layer, CdTe, that is the primary light absorber and hosts the photoconversion, is just a few μm (between 1.5 and 3). In a superstrate configuration (Fig. 3), these PV cells are manufactured by depositing the front contact i.e., a transparent conducting oxide (TCO), the intermediate layer, also called buffer, and the semiconducting thin film of CdTe on a glass superstrate (NREL, n.d.). The absorber layer is deposited through physical vapor deposition onto the stack of other layers: this process consists of sublimed CdTe which is transported to the substrate by an inert gas (Silveira Camargo et al., 2024). The TCO is transparent to visible light and highly conductive and can be made of SnO_2 or Cd_2SnO_4; the buffer layer of cadmium sulphide (CdS) improves electrical properties between TCO and CdTe (NREL, n.d.). The back layer can consist of copper/aluminium, copper/graphite or graphite doped with copper (IRENA & IEA-PVPS, 2016). Also in CdTe thin-film panels the EVA is used as encapsulant of the PV cell and is placed after the electrical contact layers, made up of a stack of conductive metals (Silveira Camargo et al., 2024). CdTe panels are generally frameless, which helps to reduce the total weight and cost but also improves the electrical isolation avoiding unwanted conduction paths for leakage currents (Maani et al., 2020).

CdTe PV cells reached record efficiencies in laboratory (up to 22.3%) and commercially available modules show an average of around 18%. In terms of manufacturing cost, it is the only thin-film technology competitive to silicon-based technology

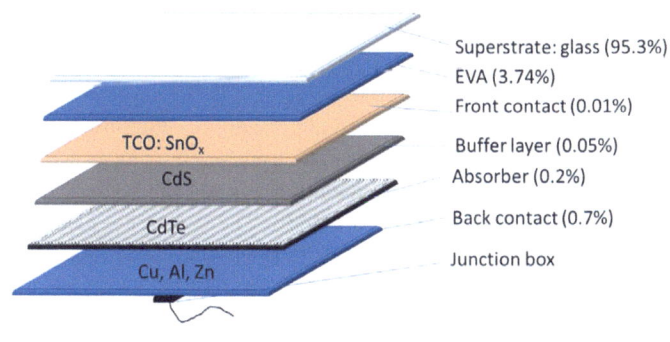

Cd;0.1%, Cu;0.6%, Al;0.05%, Te;0.1%, Zn; 1.8E-7%

Fig. 3 Cadmium Telluride (CdTe) photovoltaic cell structure and average composition by weight. Reprinted from Maani et al. (2020) with Elsevier permission

(Silveira Camargo et al., 2024). Moreover, an updated review on CdTe PV sustainability (Fthenakis et al., 2020) observed that all comparative LCA studies show CdTe PV as advantageous compared to c-Si from the environmental point of view across all selected environmental impact categories, since this technology is less energy-intensive during manufacturing.

However, the technical drawbacks of CdTe technology are cadmium toxicity and tellurium availability, being a rare metal. Silveira Camargo et al. (2024) report that 40% of tellurium global production is employed for the manufacturing of CdTe PV panels. Also, it can be noted that tellurium and cadmium are not primarily extracted but recovered as byproducts of the refinement process of other ores e.g., zinc for cadmium and copper or gold for tellurium (Silveira Camargo et al., 2024). Therefore, the role of recycling gains importance especially for Te recovery. Instead, cadmium poses health and environmental risks due to its toxicity, but it is less dangerous in the form of CdTe. In the past there was a probable overestimation of potential Cd emission from CdTe panel in case of unproper handling of the EoL module e.g., when broken modules are disposed of in landfill (Silveira Camargo et al., 2024). However, the studies considered grinded and/or unencapsulated modules which is not the common status of EoL PV panels. On the other hand, it has been observed that there is some leaching potential under aerobic acidic conditions. However, the "glass/adhesive laminate/glass" structure effectively limited this potential, reducing the leaching risk from negligible to insignificant (Fthenakis et al., 2020).

Copper Indium Gallium Diselenide (CIGS). In CIGS panels the semiconductor material is a solid solution of copper indium selenide and copper gallium selenide. This compound shows high light absorption not only in the visible spectrum but also to IR and UV radiations (Wang et al., 2024a). In fact, CIGS conversion efficiency is over 23% and is therefore the highest among thin-film technologies (Wang et al., 2024a). Additionally, CIGS solar cells show high resistance to radiations (Jean et al., 2015) and can be placed on flexible and lightweight substrates, making them suitable for space applications (Li et al., 2023).

As regards the structure, the thin layer of CIGS absorber is deposited on a back contact made of molybdenum, which is then deposited on a substrate, typically made of glass, but also steel or polymeric foils can be used (IRENA & IEA-PVPS, 2016; Wang et al., 2024a). On top of the semiconductor a buffer layer is needed to form the junction for the PV effect and is normally made of cadmium sulphide (CdS), but also Cd-free alternatives are available. The transparent front contact is made of zinc oxide (ZnO) or other transparent conducting oxides (TCOs) finished by an anti-reflection coating (IRENA & IEA-PVPS, 2016). The functional layer and the Mo layer are encapsulated in the EVA film and the sandwich structure is completed by a glass superstrate, protecting the underneath layers from oxidation and degradation (IRENA & IEA-PVPS, 2016; Wang et al., 2024a). The typical structure is illustrated in Fig. 4.

CIGS cell composition by weight can be summarised as 89% glass, 7% aluminium, 4% polymers, and less than 1% is composed of semiconductor and other metals (including 10% copper, 28% indium, 10% gallium, and 52% selenium) (IRENA & IEA-PVPS, 2016).

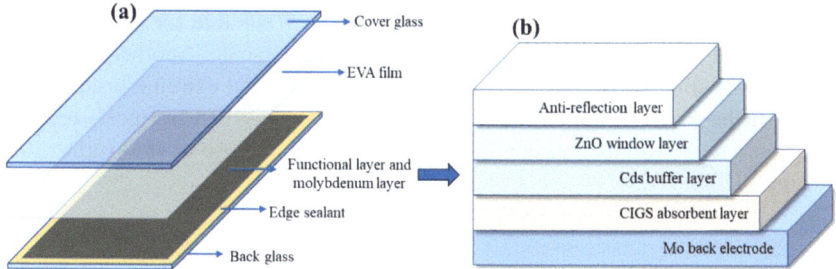

Fig. 4 Copper Indium Gallium Diselenide (CIGS) photovoltaic cell structure. Reprinted from Wang et al. (2024a) with permission from Elsevier

Since the semiconductor compound shows good light absorption properties, the thickness of the active layer is reduced to 1.5–3 μm (IRENA & IEA-PVPS, 2016) and CIGS PV technology requires lower amounts of raw materials compared to other alternatives (Li et al., 2023). However, some of these elements are critical raw materials (CRMs), i.e., copper and gallium, while indium has disappeared from the fifth list of CRMs as published in 2023. Nevertheless, both gallium and indium are not found in specific deposits but are instead obtained as byproducts of other metallurgical processes, like aluminium and zinc extraction, respectively. Therefore, recycling can play a pivotal role as the content of these metals in EoL CIGS PV panels is significantly higher than in the ores from which they are normally extracted i.e., bauxite for gallium and zinc ores for indium (Wang et al., 2024a). The replacement of some materials such as aluminium instead of indium, or silver instead of copper has been investigated (IRENA & IEA-PVPS, 2016). Another issue is related to the content of heavy metals like cadmium in the buffer layer that can be released into the environment if the EoL PV panel is not handled properly (Li et al., 2023), but some cadmium-free materials has been tested for potential replacement (IRENA & IEA-PVPS, 2016). Also the presence of selenium, as the metal elements in the active layer are metal selenides (Li et al., 2023), poses some risks since it can accumulate in the food chain and form hydrogen selenide, which is not only toxic but also carcinogenic (Wang et al., 2024a).

2.3 Market Shares

Nowadays c-Si modules dominate the solar PV market accounting for 95% of global solar PV capacity additions in 2020, followed by thin-film alternatives: CdTe covers about 2.5%, CIGS about 2%, and a-Si about 0.5% (Carrara et al., 2020). With respect to 2010, it was observed that c-Si had gained importance by replacing CdTe (−2.5%), a-Si (−1.5%), and CIGS (−1%), and it is expected to remain the most widespread technology even in the future. However, there is still uncertainty on which alternative technology could be achieving higher market shares: particular attention is given to CdTe and Gallium Arsenide (GaAs), which is suitable for multi-junction solar cells that have reached 50% efficiency at laboratory scale. Another alternative are perovskite solar cells, that show promising efficiency but extreme instability to moisture in the air (due to high solubility in water its thickness can be quickly reduced by exposure to environmental conditions) and a significant use of lead (up to 10% by weight) (IEA, 2021).

2.4 Material Requirements

The materials used in PV technologies can be differentiated into two main groups: general materials used in the PV modules and balance of system (i.e., structural and electric components), which can be considered common to all technologies, and the materials that are necessary for the solar cell itself, depending on the specific module type. Detailed assumptions must be made for the specific materials used in the solar cells according to the different technologies. In fact, differences in mineral intensities come primarily from differences in module types: c-Si PV panels typically contain about 5% silicon (solar cells), 1% copper (interconnectors), and less than 0.1% silver and other metals (IRENA & IEA-PVPS, 2016), by weight. Instead, thin-film technologies require more glass but less minerals compared to c-Si. CdTe panels require cadmium and tellurium, while CIGS require indium, gallium and selenium, but they do not use silicon or silver (Komoto et al., 2022).

The general materials are not expected to change significantly. These material intensities are expected to be affected by innovations only to a minor extent, in line with the assumption that PV modules will continue to be based on similar design structures: Carrara et al. (2020) assumed a maximum demand reduction of 20%. On the other hand, the intensity of materials specifically used in solar cells e.g., silicon, silver, cadmium and tellurium, copper, indium, gallium, selenium and germanium, might be more affected by technological developments and the evolution of market shares. Figure 5 summarises the average material composition by PV cell technology.

In terms of raw materials demand, following the study of Carrara et al. (2023), Table 1 is compiled, highlighting also the specific use of the identified material for the PV technology.

3 Legislative Framework

The European Union has adopted PV-specific waste regulations: in the EU Waste Electrical and Electronic Equipment (WEEE) Directive (2012/19/EU) addressing the waste management of all electrical and electronic devices at the end of life, there is the explicit inclusion of end-of-life PV modules. According to the legislation (Annexes III and IV) photovoltaic panels fall under category 4 "Large equipment".

Fig. 5 Average material composition according to the PV panel technology. Adapted from IRENA and IEA-PVPS (2016)

Table 1 Raw materials employed for photovoltaic technologies (Carrara et al., 2023). *Note* Critical and strategic raw materials are underlined

Use	Material
c-Si solar cells	<u>Silicon</u> as semiconductor materials in crystalline solar cells <u>Silver</u> as conductive paste for crystalline solar cells <u>Boron</u> as dopant for silicon-based wafers <u>Gallium</u> as dopant in semiconductors
CdTe solar cells	Tellurium and cadmium in active material
CIGS solar cells	<u>Gallium</u> as dopant in solar cells <u>Copper</u> in semiconductor compound in solar cells Selenium, indium, tin in solar cells as indium-tin-oxide Molybdenum as back contact
PV module structure	<u>Aluminium</u> in frames Iron and <u>nickel</u> in steel and stainless steel
Electric cables	<u>Copper</u> in cables, wires, inverters Lead in alloys with tin for soldering circuits and interconnectors

The WEEE Directive sets gradually higher targets concerning collection, preparation for reuse, recycling and recovery rates. From 2018, the Member States (MS) are required to reach a collection rate of all WEEE at least of 85% of the WEEE generated in the MS or, as an alternative, 65% of "the average weight of EEE placed on the market in the three preceding years" in the MS (Article 7). With respect to WEEE belonging to category 4, from 15th August 2018 and beyond, the required targets are (Article 11 and Annex V): recovery rate of 85% and preparation for reuse and recycling rate of 80%.

Similarly to other WEEE, the extended producer responsibility (EPR) principle is applied also to PV panels. This means that the organisation of collection, treatment and financing at the end of life is responsibility of those who place on the European market the EEE. Producers can choose to comply with EPR obligations individually or collectively, via Producer Responsibility Organisations (PROs) and compliance schemes.

Moreover, reporting PV panels mixed with other category 4 WEEE prevents a detailed monitoring of this waste stream and can also cause distortions in the calculation proving the compliance with the targets set by the WEEE Directive, especially the collection rate. In fact, due to the long lifetime of PV panels, the amount that can be collected is not aligned with the amount put on the market and often PROs need to compensate the low return of this appliances by collecting higher amounts of other category 4 EEE (WEEE Forum, 2021). This issue raises doubts about the calculation of the target based on the weight of the EEE put on the market.

The European List of Waste (European Commission, 2000) provides a reference and common nomenclature throughout the EU in order to enhance waste management activities by the adoption of common coding of waste, providing the characteristics for the distinction between hazardous versus non-hazardous waste. For PV panels specific ELW codes can be identified:

- ELW 16.02.14: Industrial WEEE;
- ELW 16.02.13*: Discarded equipment containing hazardous components;
- ELW 20.01.36: Municipal waste, used EEE;
- ELW 20.01.35*: Discarded EEE containing hazardous components.

In special cases, e.g., amorphous silicon, also the code 17.02.02 (Construction and demolition waste glass) is used.

However, the classification of waste PV panels still lacks harmonisation and specificity: they may be considered hazardous, non-hazardous or both; they may be classified as household, non-household or both; in some Member States they may be reported separately in a dedicated WEEE category, or in category 4 or as part of the latter in category 4b (as required by Article 2.2 of Decision (EU) 2019/2193) (WEEE Forum, 2021).

4 End-of-Life Management Practices in a Circular Approach

Over the past decade, recycling technologies for PV modules have been studied and developed to a considerable extent as shown by the patent analysis performed by IEA PSPV Task 12[1] (Komoto et al., 2018; Wambach et al., 2024). However, many technologies are not entirely commercialized, nor have achieved high levels of material recovery, not even for crystalline silicon PV technology, which is the dominant on the market. The most widespread recycling technology is the mechanical

---

[1] The IEA Photovoltaic Power Systems Programme (PVPS) bring together groups of experts and researchers to foster international cooperation and knowledge on photovoltaic solar energy; Task 12 focusses on sustainability of PV technologies.

recycling by glass recyclers, which is applied especially to c-Si panels, due to the predominant presence of laminated glass in their composition (Komoto et al., 2022). However, specific processes have been developed in order to enhance the recovery of materials other from glass.

In general, PV panel waste recycling can be divided in three steps:

(1) pre-disassembly;
(2) delamination i.e., processes to eliminate encapsulant from laminated structures;
(3) processes to recover metals and glass.

In the following paragraphs the recycling methods will be presented according to the solar cell technology.

4.1 Recycling of c-Si PV Panels

So far, EoL PV panels have been a waste stream generated in a variable and relatively minor amount, thus preventing the development of dedicated recycling facilities. In fact, PV modules waste have been mostly treated with mechanical processes by glass recyclers that can adapt the existing treatment for flat glass to batches of PV modules. The target of the mechanical process is the recovery of the laminated glass in the PV module. The output of the process is a crushed glass fraction that, due to the potential contamination with silicon, polymers and metals, has restricted secondary uses (Komoto et al., 2018).

The process can be summarised in the following steps (Held & Wessendorf, 2024): (1) frame and junction box removal, either manually or automatically; (2) one or more steps of crushing by using shredders or hammer mills; (3) manual sorting; (4) ferrous metal separation by magnetic separators and non-ferrous metals separation by eddy-current; (5) screening to separate glass and polymer film pieces.

The refinement of the glass fraction can occur by relying on typical processes of glass sorting lines: optical sorting techniques can be applied to remove impurities such as stones, porcelain and ceramics, but also to separate polymeric foil with solar cells from the crushed glass fraction, which can also be sorted by colour (Wambach et al., 2017). The typical process of laminated glass recycling is shown in Fig. 6.

The main components, such as glass, aluminium and copper, are recovered, resulting in cumulative yield of more than 85% (Wambach et al., 2017). Energy consumption is reported between 46 and 84 kWh/t based on Maltha BE recycling plants, a laminated glass recycler operating in Belgium, the Netherlands, and Portugal (Wambach et al., 2017), and Reiling operating glass recycling plants processing c-Si PV panels in Germany (Wambach et al., 2024).

The advantage of mechanical operation in laminated glass recycling facilities is its low-cost and the reduced need for additional investments (Komoto et al., 2018). Moreover, the mechanical approach allows to reach high recovery rates in term of modules mass: being around 85%, it is compliant with the recovery rate required by the WEEE Directive. However, the recovery is still limited in terms of value, since

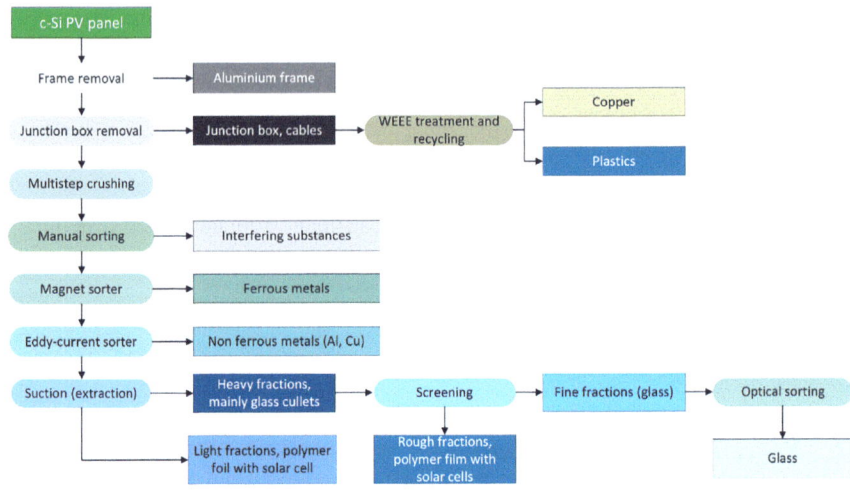

Fig. 6 Schematisation of c-Si PV panels recycling by laminated glass plant. Adapted from Held and Wessendorf (2024)

high-value materials such as silver and silicon, which are present in small percentage, cannot be recovered by this process. In fact, the relative material value of a c-Si PV panel is represented mostly by silver (47%) and aluminium (26%), followed by silicon (11%), copper (8%) and glass (8%) (IRENA & IEA-PVPS, 2016).

Considering the market value of the materials contained in c-Si PV modules, the R&D activities have been drawing the attention on the recovery of other components especially silver (representing 47% of the total value) and silicon (11% of the total value), which also result to be precious or critical raw materials. Moreover, the crushed glass fraction is significantly contaminated by silicon, polymers and metals and therefore it is suitable only for secondary applications such as thermal insulating material in the glass-foam or for glass-fibre manufacturing (Komoto et al., 2018). Due to these reasons, mechanical recycling by laminated glass recyclers is often considered as downcycling, while advanced recycling technologies focus on the recovery of materials in wafers or thin-film cells (Deng et al., 2019; Mao et al., 2024).

The following paragraphs describe in detail the three main steps of advanced recycling technologies.

Advanced c-Si Recycling step (1) Pre-disassembly. The pre-treatment phase comprises the removal of the metal frames, generally made of aluminium, and of junction box and cables. The dismantling can be manual or made by automated systems e.g., in the FRELP (Full Recovery End-of-Life Photovoltaic) project the cutting and tearing of aluminium frame is automated as well as the detachment of cables are operated by a mechanical arm (Latunussa et al., 2016a, 2016b). The disassembled elements (frame, cables and junction box) can be directly recovered with rates above 90% (Ardente et al., 2019).

Advanced c-Si Recycling step (2) Delamination. The most difficult step is the delamination i.e., eliminating the encapsulant from the laminated structures. The elimination of the encapsulant can be performed by thermal (pyrolysis, EVA combustion), mechanical (shredding or milling, hot knife, vibrating knife) or chemical (dissolution of EVA by inorganic or organic solvents (+ultrasonic radiation or microwaves)) processes; the advantage of thermal or chemical technologies is that the recovery of glass and Si-cells can happen without breakage.

Mechanical delamination typically relies on cutting the encapsulation layer, shredding or milling as well as crushing and scraping the glass and the other layers (Komoto et al., 2018). In the PHOTOLIFE project two mechanical approaches based on automatic shredding were tested: (a) simple crushing method using a two-bladed rotor and (b) crushing by a two-bladed rotor followed by hammer milling. Comparing the size distribution of the crushed pieces the conclusion was that the second option performs better and, according to the size class of the crushed pieces, it was observed that different further treatments could be applied: those with diameter $d > 1$ mm could undergo combustion at 650 °C to separate the polymers, those with $0.08 < d < 1$ mm could be recovered as glass, and those with $d < 0.08$ mm could be treated by a hydrometallurgical process to recover metals (Komoto et al., 2018);

NPC Inc. and Hamada Corporation developed a different delamination method based on a heated cutter for cutting the encapsulation layer alongside the cover-glass. The cutter is inserted into the bonding plane between the glass and the encapsulation layer while avoiding damage to the glass surface (Komoto et al., 2018). Similarly, the European project FRELP resulted in the development of an automated machinery for thermo-mechanical delamination. The laminated structure is heated by an infrared heater (to 90–120 °C) in order to soften the encapsulation layer; then the module is inserted into a roller mill equipped with a vibrating knife which performs the delamination of the glass from the rest of the structure i.e., the EVA film, the Si-cell, the electrodes and the backsheet, which are further treated (Komoto et al., 2018). Other possible methods are:

- high voltage fragmentation, i.e. applying a high voltage pulse discharge while the material is immersed in water to achieve its crushing, which has been tested with the aim of improving the recovery of metals such as silver, tin, copper, silicon, and aluminium from PV panels; it was observed that tuning the process parameters e.g. pulse rate and field strength, could lead to the concentration of metals in specific size fractions and of silicon with high purity in others (Nevala et al., 2019; Preet & Smith, 2024);
- laser application, consisting in the irradiation by an optical fibre pulsed laser from the back of the PV cells, has been proposed to separate the EVA layer without damaging the cells; the irradiation is followed by the mechanical removal of the EVA from the solar cell by peeling (Li et al., 2022; Preet & Smith, 2024).

The main drawback of mechanical approaches is that they often prevent the intact recycling of silicon wafers. While breakage is obvious with shredding, it also occurs when cutting the encapsulation layer and scraping non-glass layers, leaving only the glass recoverable without damage.

The recovered product is generally addressed as "contaminated glass cullets" (Deng et al., 2019; Mao et al., 2024) as it is mainly composed of glass cullet, but a mixture of silicon, metals and plastic components cannot be completely separated.

Another option is to apply thermal processes which are generally based on combustion which causes the decomposition of the encapsulation layer. Typically, PV modules are heated in a furnace at temperature ranging between 500 and 600 °C with heating in two stages so that polymeric components are burned and the remaining materials (e.g., Si cells, glass and metals) can be separated (Komoto et al., 2018).

As an alternative to combustion pyrolysis can be applied in order to decompose EVA. Generally, operating conditions require an inert atmosphere (e.g., nitrogen), temperatures about 500 °C and a retention time of 30 min to 1 h (Deng et al., 2019). The pyrolytic delamination can be operated also in a fluidized bed reactor (Mulazzani et al., 2022).

One general advantage of the thermal approach with respect to the mechanical process is that the recovery of glass and Si cells without damage can be achieved (Komoto et al., 2018). However, cracking has been observed and attributed to the mechanical stress given by the gas generated by EVA decomposition and trapped behind the glass as well as by the thermal deformation of the EVA on the back (Deng et al., 2019).

The advantage of pyrolysis compared to combustion is that the removal of EVA is enhanced and almost no residues are left; however, it is a slower process and the thermal stress can cause the breakage of the glass (Maani et al., 2020). On the other hand, the thermal approach is generally energy-intensive and therefore reaching its economic viability requires the treatment of higher quantities (Komoto et al., 2018; Preet & Smith, 2024) and the internal energy recovery from EVA combustion or from combustible gasses generated by EVA decomposition in pyrolytic conditions (Deng et al., 2019; Mao et al., 2024).

Another issue related to thermal processes is the generation of flue gases that need to be treated. Especially the presence of fluorinated compounds in the polymeric compounds of the backsheet generates fluoride gases that may cause damage to the furnace and require pollution control measures e.g., flue gas treatment lines (Komoto et al., 2018). In some cases a two-stage process is applied to overcome this issue: the PV panel is firstly heated at lower temperatures (150 or 330 °C) in order to achieve the softening of EVA and to separate the backsheet mechanically or manually; then the second step reaches higher temperatures and the EVA is combusted (at 400 °C) or pyrolyzed (at 500 °C) in order to recover glass and PV cells (Preet & Smith, 2024; Wang et al., 2022). Ardente et al. (2019) analysed the generation of fluorinated compounds (e.g., HF) during the combustion process and their impacts from an LCA point of view, focussing especially on acidification and human toxicity. In the study, the comparison between three different backsheets is provided: two fluorinated plastics i.e., Tedlar[2]-PET-Tedlar (TPT) and Kynar[3]-PET-Kynar (KPK), in addition to a fluorine-free backsheet i.e., PET-PET-EVA (PPE). In the assessment of the

---

[2] Polyvinyl fluoride (PVF).

[3] Polyvinylidene fluoride or polyvinylidene difluoride (PVDF).

incineration, the use of energy and reagents for the abatement of acid gasses was also included. The results show that in most impact categories the fluorine-free alternative has lower impacts, while the scenario with KPK backsheet determines the highest impacts due to the higher fluorine content.

Also, chemical processes which imply the immersion of PV modules in either organic or inorganic solvents in order to achieve the separation of different components by chemical dissolution, can be applied to achieve the delamination. The most investigated inorganic solvent is nitric acid (HNO_3) which requires 24 h for the dissolution of EVA but generates toxic gaseous emissions (Deng et al., 2019; Preet & Smith, 2024). Also a KOH-ethanol solution has been tested (Preet & Smith, 2024). Among organics solvents, trichloroethylene (TCE) succeeds in the delamination without damaging the c-Si PV cells, but requires a temperature of 80 °C and a reaction time of 10 days (Deng et al., 2019). Microwaves or ultrasonic radiation can be used to improve the speed of the process (Deng et al., 2019).

Generally, chemical methods show the advantage of recovering materials with the highest purity and recovered Si cell are not damaged. However, treatment times are longer and processes have high costs due to chemical reagents and require the proper management of liquid waste (Komoto et al., 2018; Mao et al., 2024).

Advanced c-Si Recycling step (3) Recovery of Silicon and Metals. The third step allows the recovery of silicon and other metals contained in the solar cells, specifically in the electrodes and ribbons, which include silver, copper and aluminium.

Recovering silicon can be achieved by chemical approaches such as etching or leaching, which can be applied to intact wafers but also to solar cell powders or ashes (Mao et al., 2024; Wang et al., 2022). The purity of the recovered silicon can differ significantly depending on the undergone treatments but also on the conditions of the final etching process e.g., composition of the solution, temperature, treatment duration (Preet & Smith, 2024). Two main categories of recovered silicon can be distinguished (Preet & Smith, 2024; Wang et al., 2022):

- low-grade silicon powder, which can be used in anode materials for lithium-ion batteries and requires the conversion of the recovered silicon into nano-scale or porous silicon powder;
- silicon wafer with high-purity and a certain thickness, which could be reused in the PV industry.

Additionally, the recovery of other metals from the leaching solution can be performed by hydrometallurgical or pyrometallurgical processes (Mao et al., 2024).

Etching involves the use of chemical solutions to remove specific layers or components from the silicon wafers. With the aim of recovering clean Si wafers, precise etching conditions should be applied aiming at the removal of anti-reflective coating, silver or aluminium electrodes, aluminium coating and n-p junction (Deng et al., 2019; Preet & Smith, 2024). Etching of c-Si PV cells requires both acid and alkali agents and can be distinguished by the presence or absence of hydrofluoric acid (HF) that is highly toxic and corrosive and therefore its substitution has been tested in some research (Mao et al., 2024; Wang et al., 2022). A number of etchants and combinations have been reported in literature (Preet & Smith, 2024; X. Wang et al.,

2022) and some researchers (Klugmann-Radziemska & Ostrowski, 2010) have tried to define a universal etching process for c-Si cells but they concluded that the compositions of etching solutions need to be adjusted according to the type of PV cells to be recycled. However, some compounds appear to be commonly employed. The most common acid reagents include nitric acid (HNO_3) and hydrofluoric acid (HF) which can be substituted by phosphoric acid (H_3PO_4).

Leaching, similarly, uses chemical solutions to dissolve metals from the PV panel components i.e., extract specific components or metals from a solid material into a liquid medium. Mostly acid solvents are used to leach metals like silver and copper but also tin and lead from the PV cell. Common acid leaching agents include nitric acid (HNO_3) targeting copper recovery and sulphuric acid (H_2SO_4) used to recover silver (Preet & Smith, 2024; Wang et al., 2022).

After leaching, metals need to be extracted from the leaching solution and purified. Metal extraction can rely on hydrometallurgy and pyrometallurgy (Mao et al., 2024). It was observed that most research focuses on silver recovery (Deng et al., 2019) probably due to its value (about 47%).

Hydrometallurgy involves the use of aqueous chemistry to recover metals (Deng et al., 2019; Lee et al., 2013). Some common hydrometallurgical methods are:

- Precipitation used to recover metals from the leaching solution. By precipitation, chemical reactions are used to convert dissolved metals in the solution into solid precipitates. This is often achieved by adding specific reagents or changing the conditions of the solution, such as pH or temperature. The precipitated metals can then be separated from the solution by filtration or centrifugation. For example, silver can be recovered from the leaching solution using HCl as precipitating agent.
- Electrochemical processes using electric currents to extract metals from leaching solutions. For example, electrolysis involves passing an electric current through a solution containing ions, causing them to undergo chemical reactions at the electrodes. In the context of PV panel recycling, electrolysis can be used to further refine and purify the recovered metals obtained through leaching and precipitation.
- Metal replacement commonly used for the recovery of metals from leaching solutions exploiting a more reactive metal to displace a less reactive metal from its aqueous solution. In the case of silver, zinc powder has been used.

When applying hydrometallurgical processes, it is fundamental to consider the generation of liquid waste that must be treated before the discharge. The leachate needs to be neutralised and in case of acid leaching solutions using alkali for the neutralisation allows also the extraction of metal elements still present in the solution as metal ions can combine with hydroxide to precipitate (Mao et al., 2024).

Pyrometallurgy, instead, involves high-temperature processes (such as smelting and roasting) to extract and purify metals. The application of pyrometallurgical processes to EoL PV cells is still limited but recent research has tested the recovery of silicon, aluminium, and silver. Nevertheless, reaching high temperatures significantly increases the energy consumption and can lead to the generation of toxic gases which need adequate treatment (Mao et al., 2024). Figure 7 summarises graphically the options to manage EoL c-Si PV panels.

4.2 Recycling of CdTe PV Panels

In the case of thin-film PV panels, the semiconductive material is generally deposited on a substrate and encapsulated: therefore, also this configuration requires the removal of the EVA encapsulant layer to allow the recovery of the solar cells' components and especially of the semiconductor. As regards advanced recycling three main steps can be distinguished: (1) pre-disassembly; (2) delamination or comminution by thermal, mechanical or chemical processes; (3) recovery of metals i.e., cadmium and tellurium through materials separation, metal dissolution and solid–liquid separation (Silveira Camargo et al., 2024).

Advanced CdTe Recycling step (1) Pre-disassembly. The pre-treatment of CdTe consists in the disassembly of cables and junction box while there is no need for the frame removal since CdTe panels are often frameless (Maani et al., 2020).

Advanced CdTe Recycling step (2) Delamination. Shredding and milling can be applied to reduce the size of the thin-film laminate for further treatments; these processes are suitable for both complete and broken PV modules (Berger et al., 2010). The outputs are glass cullet of different sizes (below 20 mm) still laminated with the semiconductive layer and EVA foil fragments (Berger et al., 2010). Nevertheless, this process has demonstrated not being able to remove EVA completely from glass and semiconductor and therefore a further treatment is required. The crushed materials usually undergo attrition as following material separation treatment (Maani et al., 2020; Silveira Camargo et al., 2024).

Alternatively, delamination by thermal treatment has been tested by heating CdTe panels at 500 °C in a lab-scale furnace (Berger et al., 2010; Maani et al., 2020). Thermal delamination allows a complete EVA removal and has affordable costs, but is energy intensive, generates emissions and can damage the cells since high temperature is causing defects and cells degradation (Maani et al., 2020; Silveira Camargo et al., 2024). On the other hand, it results in a clean back glass and the superstrate glass with the semiconductor that can be subjected to vacuum blasting (i.e., a material separation method).

Chemical delamination can be based on nitric acid dissolution, leading to a complete removal of EVA and metal coating on the wafer. However, this inorganic acid can damage the cells and generate harmful emissions and waste (Maani et al., 2020). Also, the dissolution by organic solvent shows promising results on EVA and glass recovery but can be time-consuming and generate harmful emissions (Silveira Camargo et al., 2024).

Advanced CdTe Recycling step (3a) Materials separation. The aim of this step is to separate glass and semiconductor materials; also EVA is targeted in case it has not been chemically dissolved or burned by thermal treatment.

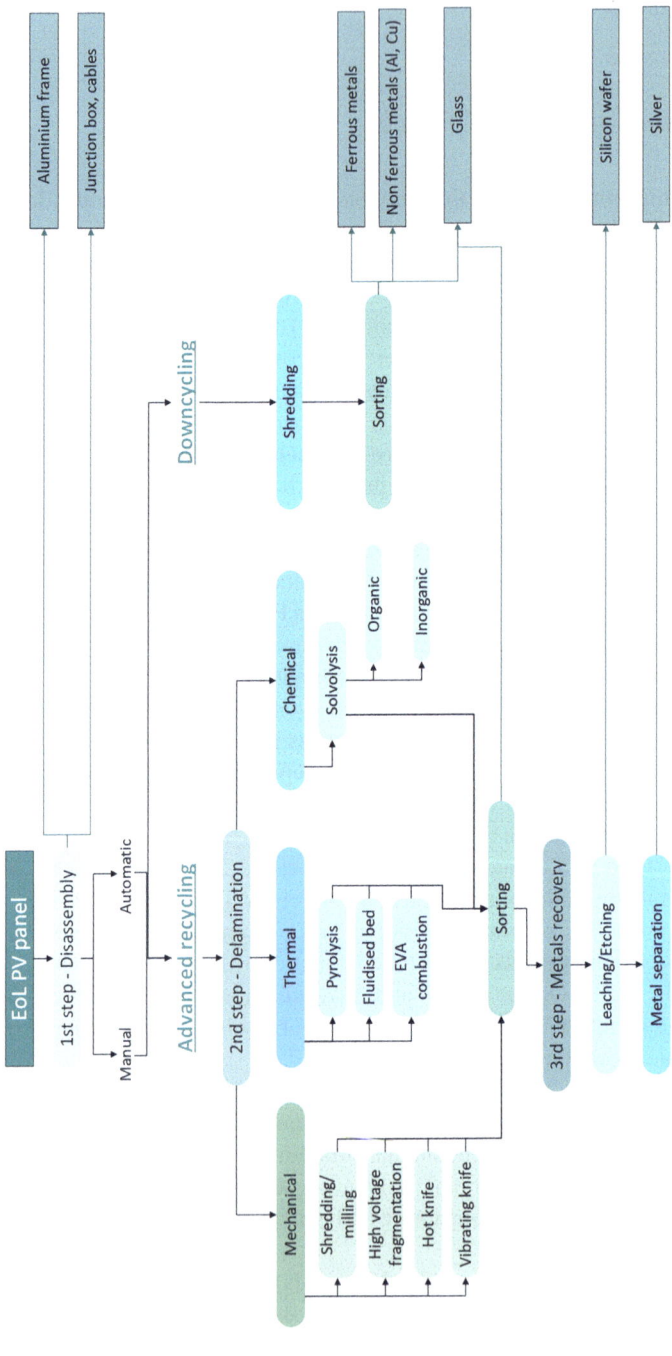

Fig. 7 Possible recycling technologies for EoL crystalline silicon solar PV panels. Adapted from Wang et al. (2022)

Attrition is normally applied to crushed laminates with the aim of further separating the materials (Silveira Camargo et al., 2024). It is a wet mechanical process which exploits shear and friction forces generated on the surface of particles for the separation and avoids the use of chemicals (Maani et al., 2020). Attrition is generally followed by sieving for separating glass and EVA with size above 150 μm from glass and semiconductor materials smaller than 150 μm. The fine fraction is then sent to a flotation process (Maani et al., 2020). The main drawback of flotation is that requiring rinsing and sieving, significant amounts of valuable materials can be lost (Silveira Camargo et al., 2024).

As an alternative, vacuum blasting can be applied: in this process an abrasive blast medium (e.g., iron powder, glass beads, aluminium oxide) strikes the surface of the material and removes the thin-film layer. By creating vacuum on the surface, the blasting medium and the abraded material is removed by suction, preventing dust emissions and allowing a close circuit for the blasting material (Berger et al., 2010). Its main advantages are avoiding the use of chemicals and recovering clean glass; however, the process is slow and generally requires additional post-treatments, either mechanical or chemical (Silveira Camargo et al., 2024).

Advanced CdTe Recycling step (3b) Metals extraction and solid–liquid separation. According to Maani et al. (2020) there are five different material extraction methods reported in literature: electrolysis, ion exchange and electrowinning, liquid–liquid extraction, precipitation and oxidation reduction. However, only leaching is described in other publications (Berger et al., 2010; Silveira Camargo et al., 2024). In fact, the recovery of metals contained in CdTe PV panels can be achieved by leaching with specific solvents. Especially nitric acid leads to very high tellurium recovery (96%) but does not allow cadmium dissolution. Instead, sulphuric acid (H_2SO_4) and hydrogen peroxide (H_2O_2) are coupled to leach both cadmium and tellurium. Microwaves or ultrasounds can be also used to speed up the process (Silveira Camargo et al., 2024).

After the dissolution, metals should be recovered from the solvent: precipitation using sodium hydroxide (NaOH) can be used to recover tellurium and cadmium in a solid form. Also, ion-exchange resins can be used to remove the metal ions from a solution (Silveira Camargo et al., 2024).

Industrial application. Currently, the most widely known recycling process for CdTe available at industrial scale is the one by First Solar, which is also the largest CdTe panels manufacturer (Silveira Camargo et al., 2024). The method consists in a combination of mechanical and chemical processes (Fig. 8).

The First Solar recycling process steps are the following (Held & Wessendorf, 2024; Silveira Camargo et al., 2024):

1. removal of cables and junction box;
2. mechanical comminution down to 4–5 mm by two-stage crushing using a shredder followed by a hammer mill: this allows the creation of a large edge area for the subsequent etching process;
3. removal of semiconductor: etching process for breaking the laminated foil made of EVA and glass and to achieve the dissolution of the semiconductor material.

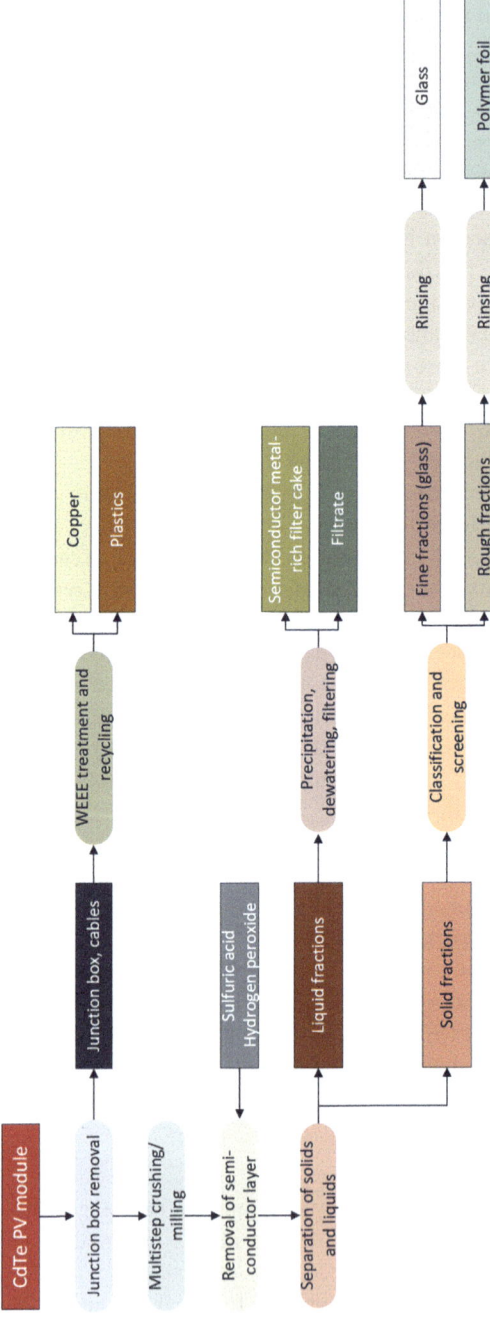

Fig. 8 Schematisation of CdTe PV panels recycling process by First Solar. Adapted from Held and Wessendorf (2024)

Dust is extracted via a ventilation system equipped with a high-efficiency particle filter.
4. dissolution of CdTe semiconductor layer: acid-leaching in oxidative conditions using a mixture of sulphuric acid and hydrogen peroxide (H_2SO_4 and H_2O_2) happening in a rotating leach drum for 4 to 6 h;
5. separation of solids and liquids:
   - after the drainage of the drum, the liquid containing cadmium and tellurium (Cd^{2+} and Te^{4+}) undergoes precipitation: the metals are precipitated from the acid by raising the pH value using NaOH; the precipitated metals can be found in form of a sludge that is thickened e.g. by filtering and resulting in Cd and Te concentrated in a filter cake;
   - the remaining solid fraction is separated using a classifier and a vibrating screen so that EVA flakes concentrate in the coarse fraction while glass cullet in the fine fraction;
   - both solid fractions are washed with water in order to rinse the residual process medium.
6. It has to be noted that he recovered glass cullet is still restricted to secondary applications with lower quality requirements e.g. foam glass production.

4.3 Recycling of CIGS PV Panels

The recycling of CIGS PV panels involves separating the different components also to recover the cover glass, followed by treatments to enrich, separate, and purify the high-value metals. The glass removal normally relies on a combination of heat treatment and mechanical process in order to soften the EVA encapsulation and then remove the superstrate glass by mechanical push off and manual disassembly (Li et al., 2023). Subsequently, the solar cell that still presents some other residual layers such as EVA, ZnO window layer and CdS buffer layer undergoes leaching in acetic acid so that the unwanted components are dissolved. So, the CIGS layer, still connected to the substrate glass, is removed by mechanical scraping and the base glass is treated again with nitric acid to remove the molybdenum coating and possible other residues. After scraping, the CIGS active layer is in the form of a powder that can be further treated, while the base glass can be reused and the cover glass can be recycled in the glass industry (Li et al., 2023). As the use of CIGS on flexible substrates shows great potential, suitable treatments for this application without base glass should be developed (Li et al., 2023).

Another option for removing the encapsulant layer is pyrolysis, similarly to other PV recovery technologies. Different conditions for processing have been tested and reported in the literature, but in general the process requires temperatures around 500 °C and has a duration within an hour (Wang et al., 2024a).

After their enrichment, valuable metals such as copper, indium, gallium and selenium are present as mixed selenides and their recovery after the pretreatment can happen by pyrometallurgy, hydrometallurgy or combined methods (Li et al., 2023).

Pyrometallurgy separation exploits high temperatures and metallurgical processes such as roasting i.e., gas–solid reactions at elevated temperatures with the goal of purifying the metal components. In case of EoL CIGS, pyrometallurgical methods are applied mostly to recover selenium (Wang et al., 2024a). In fact, oxidative roasting is applied to recover selenium as selenium dioxide (SeO_2) which is obtained by volatilization and condensation, followed by reduction to high-purity selenium (Li et al., 2023). Different reaction times, in the order of hours, and different temperatures in the range 500–1000 °C are reported in the literature (Wang et al., 2024a). The other mixed oxides, instead, undergo chloridising roasting at lower temperature (around 300 °C) utilizing chloride as a reducing. Ammonium chloride (NH_4Cl) shows better results in gallium and indium separation as chlorides i.e., separation rates above 90%, while copper remains in the roasting slag (Li et al., 2023; Wang et al., 2024a).

Hydrometallurgy separation is based on leaching, in order to convert the solids into a liquid phase, followed by separation methods such as precipitation and extraction. The occurring reactions between the leaching agent and the metals can lead to their dissolution or precipitation, enabling the following extraction (Wang et al., 2024a). Most commonly inorganic acids such as sulphuric acid (H_2SO_4), nitric acid (HNO_3) or hydrochloric acid (HCl) are used, while organic acids, e.g., citric acid (CH_3COOH), show a reduced leaching rate (Li et al., 2023). Direct or enhanced acid leaching can be distinguished: the latter employs reinforcing oxidisers (e.g., H_2O_2) and pressurisation of the process (Li et al., 2023). After leaching, precipitation can be induced by adjusting the pH since the forms of copper, indium, and gallium ions clearly differ at different pH values; in this way their gradient separation can happen. Extraction, instead, exploits the selectivity of specific extractants for given metals (Li et al., 2023).

In practice pyrometallurgical and hydrometallurgical techniques are frequently combined, depending on the characteristics of the waste and the specific metal extraction needs. In the case of CIGS, studies have shown that oxidative roasting for selenium removal notably improves the leaching rate of other metals in subsequent processes. Some combined techniques are (Li et al., 2023; Wang et al., 2024a):

- oxidative roasting followed by acid leaching with H_2SO_4, precipitation and extraction;
- oxidative roasting followed by acid leaching with HCl or HNO_3 and final extraction;
- sulphating roasting with concentrated sulphuric acid (H_2SO_4) followed by a second calcination to complete the phase transition of selenium combined to water leaching and alkaline leaching with NaOH.

Electrodeposition is another method, currently at experimental stage, which can be employed to separate and recover multiple elements from CIGS based on factors such as the metal solubility and differences in working potentials. Electrodeposition is an electrochemical process in which metal ions are reduced in an electrolyte and then there is the deposition of metal (Wang et al., 2024a). The advantage is to increase metal recovery rates while reducing the use of chemical reagents (Wang et al., 2024a).

4.4 Performance, Benefits, Drawbacks

Each method previously described has its advantages and disadvantages. Tables 2, 3 and 4 list the key observations for each one, according to the addressed PV technology.

Review of Life Cycle Assessment on EoL Management of PV Panels. To compare the recycling and recovery technologies from the environmental perspective, a literature review has been made looking for life cycle assessment (LCA) studies on the topic. The past decade has seen an increase in LCA publications on PV panels' end-of-life (EoL) stage, though comparison remains challenging due to varying modelling choices. The following section provides a selection of LCA studies on recycling of PV modules.

Table 2 Summary of possible recycling methods for c-Si PV panels and the relative advantages and disadvantages

Process step	Method	Advantages	Disadvantages
Mechanical recycling	Laminated glass recycling process	Low additional investments High recovery rates in terms of mass (mostly glass) Allows compliance with WEEE Directive	No recovery of high-value materials Limited value recovery
Delamination	Mechanical	Low cost Reduced energy consumption High throughput	Low recovery rate Low quality recovered materials i.e., glass cullet used for fibreglass production No recovery of valuable metals from the PV cells
	Thermal	High recovery rate Partially self-sustainable by energy recovery Pyrolysis leaves almost no residues compared to combustion	Energy intensive Management of flue gas required
	Chemical	Highest purity of recovery Low damages to Si cells	High costs due to reagents and liquid waste Longer reaction time than thermal methods

Table 3 Summary of possible recycling methods for CdTe PV panels and the relative advantages and disadvantages based on Maani et al. (2020)

Process step	Method	Advantages	Disadvantages
Delamination	Thermal	Full removal of EVA Possible unbroken cell recovery	Energy intensive Generation of harmful emissions
	Mechanical	Suitable for intact and broken panels	No removal of dissolved solids
Material separation and recovery	Attrition and flotation	No chemicals usage Clean glass recovery	Further chemical or mechanical processes needed to enrich recovered materials
	Leaching	Complete removal of metals from glass Allows further metals extraction	High use of chemicals Requires acid waste management

Several studies focus on c-Si PV panels. Latunussa et al., (2016a, 2016b) evaluate the recycling method developed during the FRELP project i.e., automatic dismantling, use of vibrating knife and IR heating, incineration with energy recovery of sandwich layer, chemical treatment of combustion ashes (acid leaching) and final metals recovery by vacuum filtration and electrolysis. The results show that the main hotspots of the process are the thermal treatment of the PV sandwich and the processes for metals recovery from combustion ashes. The following study by Ardente et al. (2019) expands the system boundaries, e.g., including potential benefits due to materials recovery, and includes the comparison with the baseline recycling process by glass recyclers and the evaluation of different types of backsheet. The study highlights that the impacts of the FRELP recycling process are higher than those of the baseline process due to its complexity and higher energy requirements. Nevertheless, considering also the credits for the recovered materials the FRELP process results in net environmental benefits. In addition, a pyrolysis scenario is introduced for the case of halogen-free backsheet: this scenario would lead to generally lower burdens in most impact categories than the standard FRELP process, but it is strictly related to the backsheet composition. Ansanelli et al. (2021) evaluate the thermo-mechanical-hydrometallurgical process developed in the RESIELP project (Cerchier et al., 2021), relying on primary data from pilot scale treatment including two lines of activities: recovery line from PV panels and glass reuse line. The hotspots in the PV recovery line are the electricity used for the heating treatment, the energy for the abatement system and the infrastructures. Including recycling credits leads to net environmental benefits in all categories except one, especially thanks to the avoided aluminium production. Lunardi et al. (2018) compare different c-Si waste management options including landfill, incineration (without energy recovery), reuse, thermal recycling (combustion of EVA), chemical recycling (using toluene + KOH + HNO_3, HF, CH_3COOH) and mechanical recycling (developed in FRELP project) but considering only the recovery of glass, aluminium and silicon. The study concludes that

Table 4 Summary of possible recycling methods for CIGS PV panels and the relative advantages and disadvantages, based on Li et al. (2023) and Wang et al. (2024a)

Process step	Method	Advantages	Disadvantages
Delamination	Thermo-mechanical (heat treatment + push off and scraping)	Relatively simple	Dissolution must be sufficient Scraping separation equipment should have high precision
	Crushing and sorting	High processing throughput	Limits high product purity Unfavourable for following metal recovery
	Pyrolysis	Efficient removal of EVA Beneficial for following metal recovery	Generation of toxic gases
Materials recovery	Pyrometallurgy	Good separation especially of selenium	Energy intensive Volatile and toxic reagents and intermediate products; requires flue gas cleaning Need for anti-corrosion equipment Currently at lab scale
	Hydrometallurgy	Simple operation Low energy consumption High recovery rate	High consumption of reagents Requires wastewater treatment Currently at lab scale
	Combined methods	High recovery efficiency for Se, Cu, In and Ga	Production of organic waste Requires flue gas cleaning Currently at lab scale

recycling methods show lower environmental impacts compared to other disposal options and chemical recycling results to be the least burdensome. However, concerns are raised due to the toxicity of substances employed during the chemical recycling treatment. In the review by Mao et al. (2024) the LCA results of c-Si panels recycling are analysed and compared when possible. The results show that incineration followed by landfilling of non-combustible components is the worse solution. Also, mechanical recycling is performing well and it is the best option in the human toxicity category. Among thermal treatments, pyrolysis generally has lower impacts than incineration. Duflou et al. (2018) report the LCA analysis of an innovative selective delamination technology for c-Si based on mechanical process compared to the baseline mechanical technology (size-reduction followed by magnetic and eddy-current separators) and to pyrolysis and acid leaching scenario. The innovative delamination, i.e., mechanical cleaving possibly assisted by a temperature increase, shows the lowest impacts in most categories, while in a few cases the thermo-chemical process results in even lower burdens. (Stolz et al., 2017).

Recently, also the LCA of the delamination process of c-Si PV cells according to hot knife technology based on primary data by NPC Inc has been published (Frischknecht et al., 2023).

Regarding CdTe PV panels, Held (2009), Held and Ilg (2011), and Sinha et al. (2012) are widely cited analyses based on First Solar recycling process. The analyses show that the required energy and auxiliaries of the module recycling process are the main contributors in all considered environmental impact categories. However, the inclusion of materials recycling credits shows that the benefits outweigh the impacts of the recycling process. Moreover, it is noted that the EoL stage has much lower impacts than the production phase (Held, 2009; Held & Ilg, 2011). Sinha et al. (2012) include in the analysis also the refining of the semiconductor material assessing its contribution to the EoL impacts. Contreras Lisperguer et al. (2020) compared the recycling of c-Si and CdTe PV panels. The study concludes that CdTe closed-loop recycling shows generally lower environmental impacts across all Recipe impact categories with few exceptions (freshwater and marine ecotoxicity and human toxicity). Another consideration is that c-Si PV recycling and silicon recovery are less well-established processes compared to the CdTe recycling process developed by First Solar, which in the study outperforms the PV recycling process in almost all categories. Maani et al. (2020) analyse the environmental impacts of different recycling methods for c-Si and CdTe PV panels. For c-Si several EoL options are included, while regarding CdTe panels only mechanical and thermal processes are assessed. The analysis reports that for c-Si the worst performance is shown by leaching with nitric acid followed by pyrolysis. Nevertheless, the study points out that generally recovering the materials of c-Si PV panels through recycling is more environmentally impactful than the corresponding virgin materials extraction. For CdTe panels, instead, both recycling methods produce very low impacts compared to panel manufacturing and especially thermal treatment is the most environmentally friendly. Also including the material separation step (floatation, attrition and leaching) the burdens remain low.

The number of publications on CIGS PV recycling is much more limited. Wang et al. (2024a) provide a brief review of LCA studies focussing on CIGS PV panels. In Rocchetti and Beolchini (2015), recycling of CdTe and CIGS panels is compared including conventional and innovative options. Conventional recycling consists of mechanical treatment, EVA sent to energy recovery and residues landfilling. Innovative methods, instead, include metals recovery (i.e., selenium, indium and gallium from the CIGS panels, and tellurium from the CdTe panels). Conventional methods result in comparable impacts for both the typologies of PV panels, while innovative ones result to be more advantageous than conventional techniques for CdTe panels due to the environmental credits associated with the recovery of valuable materials;

instead, for CIGS recycling the use of reagents is a hotspot. Nevertheless, landfilling shows higher environmental impacts than recycling in most of the analysed impact categories. Other studies focus on specific treatments, e.g., Amato and Beolchini (2019) quantify the carbon footprint of various acid leaching processes and points out that the combination of citric acid and H_2O_2 has the lowest impacts on climate change.

5 Industrial Plants and Initiatives

Fully understanding the development of the photovoltaic panel recycling sector in Europe can be quite challenging. Since EoL PV panels gain attention as a growing waste stream and the European PV waste market expands significantly, the landscape of industrial facilities managing this WEEE continues to evolve. While searches for recyclers accepting PV modules can yield numerous results, providing a complete overview remains difficult. Moreover, the limited availability of detailed information about their specific recycling processes hinders a comprehensive analysis of the prevalence of different recycling methods.

A recent report by Wambach et al. (2024) provides a general outlook of recyclers of EoL PV modules: 45 companies have been identified in Europe (including the UK), concentrated mostly in Germany (17) followed by Italy (7) and France (6).

It can be observed that mechanical recycling is still the most widespread option although the recovered material quality is downgraded compared to the input material. It is noteworthy that Reiling, operating glass recycling plants in Germany, has improved the purely mechanical processing to handle silicon-based PV modules, both crystalline silicon (c-Si) and amorphous silicon, and can now be considered the best available technology. The main recycled outputs are glass cullet (64% of the input), aluminium frames (11.5%), and other metals (1.4%). Reiling has a capacity of 10,000 tonnes per year and plans to scale up to 50,000 tonnes annually.

Nevertheless, also some innovative initiatives can be identified. LuxChemtech in Germany employs water-jet cleaning and a light-pulse treatment to recycle crystalline silicon and thin-film panels. Thanks to the high pressure, the water-jet allows the removal of polymers, solar cells and metals from the glass, which is recovered as highly pure and possibly intact. In the next steps, non-ferrous metals are separated from polymers, while silicon or semiconductor materials and metals, including copper, silver, indium and tin, are recovered by etching. LuxChemtech is operating a pilot plant with a target capacity of 1,000 tonnes per year.

Flaxres, another German company, uses light pulse technology for delaminating crystalline silicon and thin-film panels. The light pulse heats light absorbing materials, such as the silicon wafer, to allow the separation of the layers. Flaxres achieves a clean separation of glass, polymers, and solar cells fragments, which can be further treated for silver and silicon recovery.

In France, ROSI applies a batch pyrolysis process for polymers removal to obtain high quality glass cullet together with metals and solar cell fragments. The copper interconnectors and solar cells fragments are separated through existing mechanical methods, while a proprietary soft etching is applied to recover silicon and silver from cells. The plant has a capacity of 3,000 tonnes per year, producing high-purity silicon (ranging from 5 to 6N), silver, and clean glass (corresponding to 71% of the input). Moreover, ROSI collaborates with Envie, another French company, applying the NPC hot blade technology for delamination. Envie recovers in this way the frame, cables and junction box, and the glass; the laminated fraction, instead, is sent to ROSI. Envie has a capacity of 3,000 tonnes per year for crystalline silicon modules.

The Italian company Tialpi S.r.l. utilizes an automatic dismantling followed by thermo-mechanical process for delamination aiming at high-quality glass recovery. The detachment of glass layer from the PV sandwich is achieved by heating using infrared and mechanical detachment by a high-frequency knife button. The glass is then refined by sieving and optical separation of impurities. Tialpi is currently operating a pilot plant with a capacity of 1,000 tonnes per year, reaching a recovery rate of about 80% of the PV panel in terms of mass. However, there is still ongoing research to offer an alternative to incineration for the sandwich, i.e., achieving the delamination of silicon wafer (including EVA) from the PVF backsheet, as it has a market in the metallurgical sector, particularly in aluminium smelting. Moreover, aiming at improving the recovery in terms of valuable materials, further steps are studied for the industrialization, i.e., pyrolysis to remove EVA followed by acid leaching with nitric acid of the pyrolysis ashes, in order to let all metals dissolve except for silicon which can be then recovered by filtration, while silver and copper can be recovered from the leachate by electrolysis.

9-Tech is an Italian start-up which patented a technology to recover most of the PV modules, in terms both of mass and of value: 98% and 95%, respectively. The delamination happens in an oven (thermal process) and is based on combustion of EVA and backsheet (with energy recovery and flue gas cleaning). Subsequently a chemical treatment of cells with weak acids and ultrasounds allows the recovery of silver powder (being in a solid state, leaching is avoided). Also, silicon is recovered at metallurgical grade (2N purity). A pilot plant was used to treat 2 tonnes per day and currently 9-Tech is in the upscaling phase i.e., they started the permitting phase for an industrial plant.

The PV panel types being addressed by recyclers are primarily crystalline silicon (c-Si) due to their prevalence in the market. Reiling, Tialpi and ROSI are among the facilities focusing on such panels. Beyond mechanical treatment by glass recyclers, advanced and innovative technologies in photovoltaic panel recycling, currently in the pilot stages, are improving the quality and yield of recovered materials, improving the economic return by valuable materials recovery. These advanced methods enable the recovery of key components such as copper, silicon and silver.

For thin-film panels, which include technologies like cadmium telluride (CdTe) and CIGS, there is strong expertise by First Solar, achieving over 90% material recovery through its proprietary system and offering the treatment in different facilities around the world, including one in Germany. Moreover, some emerging recycling technologies show potential for application also to thin-film modules, though some may require specialised treatments.

6 Conclusions

Photovoltaic panels are distinguished by the employed semiconductive material i.e., the active part in light absorption: crystalline silicon (c-Si), cadmium telluride (CdTe) and copper indium gallium diselenide (CIGS). End-of-life PV panels are classified as Waste Electrical and Electronic Equipment (WEEE) under "large equipment" (category 4). Depending on their characteristics and power rating can be classified as municipal waste or industrial waste, hazardous or non-hazardous waste, preventing a proper monitoring of this waste stream and possibly causing distortions in the calculation of the targets set by the WEEE Directive.

Currently, the most common recycling method for PV panels is mechanical processing at glass recycling plants, focusing on laminated glass recovery, due to the predominant presence of such component in PV panels composition. This method achieves a good mass recovery rate (around 85%) but fails to recover high-value materials like silver and silicon, which constitute about 60% of the PV panel's value. Advanced recycling technologies have been explored and typically comprise three steps: 1. pre-disassembly; 2. delamination; 3. processes to recover valuable materials. Pre-disassembly removes metal frames, junction boxes, and cables. Delamination varies by semiconductor type but typically employs mechanical (e.g., shredding, hot knife), thermal (e.g., pyrolysis, EVA combustion), or chemical (e.g., dissolving EVA with solvents) methods. As a general trend, the goal is to remove the EVA encapsulation from the laminated structures while preserving the integrity of solar cells and avoiding contamination of glass cullet with a mix of silicon, metals, and plastics. For the recovery of silicon and metals (e.g., silver, copper, aluminium), chemical processes like etching or leaching are used, followed by hydrometallurgical or pyrometallurgical extraction. Thin-film panels (CdTe and CIGS) require slightly different processes due to a higher proportion of sub- or superstrate glass. As of 2024, in Europe there are at least 45 companies treating EoL PV panels, most of which rely on mechanical treatment focusing on glass cullet recovery. However, some initiatives are developing and testing at pilot scale advanced recycling methods, especially thermo-mechanical delamination combined by chemical processes to improve recovery both in terms of mass and value.

Competing Interests The authors have no conflicts of interest to declare that are relevant to the content of this chapter.

References

Amato, A., & Beolchini, F. (2019). End-of-Life CIGS photovoltaic panel: A source of secondary indium and gallium. *Progress in Photovoltaics: Research and Applications, 27*(3), 229–236. https://doi.org/10.1002/PIP.3082

Ansanelli, G., Fiorentino, G., Tammaro, M., & Zucaro, A. (2021). A life cycle assessment of a recovery process from End-of-Life photovoltaic panels. *Applied Energy, 290*, 116727. https://doi.org/10.1016/J.APENERGY.2021.116727

Ardente, F., Latunussa, C. E. L., & Blengini, G. A. (2019). Resource efficient recovery of critical and precious metals from waste silicon PV panel recycling. *Waste Management, 91*, 156–167. https://doi.org/10.1016/j.wasman.2019.04.059

Berger, W., Simon, F. G., Weimann, K., & Alsema, E. A. (2010). A novel approach for the recycling of thin film photovoltaic modules. *Resources, Conservation and Recycling, 54*(10), 711–718. https://doi.org/10.1016/J.RESCONREC.2009.12.001

Carrara, S., Alves Dias, P., Piazzotta, B., & Pavel, C. (2020). Raw materials demand for wind and solar PV technologies in the transition towards a decarbonised energy system. https://data.eur opa.eu, https://doi.org/10.2760/160859.

Carrara, S., Bobba, S., Blagoeva, D., Dias, A., Cavalli, P., Georgitzikis, A., et al. (2023). *Supply chain analysis and material demand forecast in strategic technologies and sectors in the EU: A foresight study*. JRC Science for Policy Report. EU CRM. https://doi.org/10.2760/334074.

Cerchier, P., Brunelli, K., Pezzato, L., Audoin, C., Rakotoniaina, J. P., Sessa, T., et al. (2021). Innovative recycling of end of life silicon PV panels: ReSiELP. *Detritus, 16*(16), 41. https://doi.org/10.31025/2611-4135/2021.15118.

Contreras Lisperguer, R., Muñoz Cerón, E., de la Casa Higueras, J., & Martín, R. D. (2020). Environmental impact assessment of crystalline solar photovoltaic panels' End-of-Life phase: Open and closed-loop material flow scenarios. *Sustainable Production and Consumption, 23*, 157–173. https://doi.org/10.1016/J.SPC.2020.05.008

Deng, R., Chang, N. L., Ouyang, Z., & Chong, C. M. (2019). A techno-economic review of silicon photovoltaic module recycling. *Renewable and Sustainable Energy Reviews, 109*, 532–550. https://doi.org/10.1016/J.RSER.2019.04.020

Duflou, J. R., Peeters, J. R., Altamirano, D., Bracquene, E., & Dewulf, W. (2018). Demanufacturing photovoltaic panels: Comparison of end-of-life treatment strategies for improved resource recovery. *CIRP Annals, 67*(1), 29–32. https://doi.org/10.1016/J.CIRP.2018.04.053

European Commission. (2000). Commission decision of 3 May 2000. *Offical Journal of the European Communities, L 226*. https://eur-lex.europa.eu/legal-content/EN/TXT/?uri=CELEX%3A0 2000D0532-20231206#tocId2.

Frischknecht, R., Komoto, K., & Doi, T. (2023). Life cycle assessment of crystalline silicon photovoltaic module delamination with hot knife technology. https://iea-pvps.org/wp-content/upl oads/2023/07/Report_IEA-PVPS_T12-25-2023_LCA-PV-Recycling-Hot-Knife.pdf.

Fthenakis, V., Athias, C., Blumenthal, A., Kulur, A., Magliozzo, J., & Ng, D. (2020). Sustainability evaluation of CdTe PV: An update. *Renewable and Sustainable Energy Reviews, 123*, 109776. https://doi.org/10.1016/J.RSER.2020.109776

Held, M., & Wessendorf, C. (2024). Status of PV module take-back and recycling in Germany. https://iea-pvps.org/key-topics/status-of-pv-module-take-back-and-recycling-in-germany/.

Held, M., & Ilg, R. (2011). Update of environmental indicators and energy payback time of CdTe PV systems in Europe. *Progress in Photovoltaics: Research and Applications, 19*(5), 614–626. https://doi.org/10.1002/PIP.1068

Held, M. (2009). Life cycle assessment of CdTe module recycling. In *4th European Photovoltaic Solar Energy Conference (EU PVSEC)*. https://www.icpds.com/assets/planning/final-environme ntal-impact-reports/campo-verde-solar/life-cycle-cdte.pdf.

IEA. (2021). The role of critical minerals in clean energy transitions. https://www.iea.org/reports/the-role-of-critical-minerals-in-clean-energy-transitions.

IRENA, & IEA-PVPS. (2016). End-of-life management: Solar photovoltaic panels. https://www.irena.org/publications/2016/Jun/End-of-life-management-Solar-Photovoltaic-Panels.

Jean, J., Brown, P. R., Jaffe, R. L., Buonassisi, T., & Bulović, V. (2015). Pathways for solar photovoltaics. *Energy & Environmental Science, 8*(4), 1200–1219. https://doi.org/10.1039/C4EE04 073B

Klugmann-Radziemska, E., & Ostrowski, P. (2010). Chemical treatment of crystalline silicon solar cells as a method of recovering pure silicon from photovoltaic modules. *Renewable Energy, 35*(8), 1751–1759. https://doi.org/10.1016/J.RENENE.2009.11.031

Komoto, K., Lee, J.-S., Zhang, J., Ravikumar, D., Sinha, P., Wade, A., et al. (2018). End of life management of photovoltaic panels trends in PV module recycling technologies. https://iea-pvps.org/key-topics/end-of-life-management-of-photovoltaic-panels-trends-in-pv-module-recycling-technologies-by-task-12/.

Komoto, K., Held, M., Agraffeil, C., Alonso-Garcia, C., Danelli, A., Lee, J.-S., et al. (2022). Status of PV module recycling in selected IEA PVPS Task12 countries. https://iea-pvps.org/key-topics/status-of-pv-module-recycling-in-selected-iea-pvps-task12-countries/.

Latunussa, C. E. L., Ardente, F., Blengini, G. A., & Mancini, L. (2016). Life cycle assessment of an innovative recycling process for crystalline silicon photovoltaic panels. *Solar Energy Materials and Solar Cells, 156*, 101–111. https://doi.org/10.1016/j.solmat.2016.03.020

Latunussa, C. E. L., Mancini, L., Blengini G.A., Ardente F., & Pennington D. (2016). Analysis of material recovery from photovoltaic panels. https://op.europa.eu/en/publication-detail/-/publication/3b3d0582-0c3a-11e6-ba9a-01aa75ed71a1/language-en.

Lee, C. H., Hung, C. E., Tsai, S. L., Popuri, S. R., & Liao, C. H. (2013). Resource recovery of scrap silicon solar battery cell. *Waste Management and Research, 31*(5), 518–524. https://doi.org/10.1177/0734242X13479433/ASSET/IMAGES/LARGE/10.1177_0734242X13479433-FIG7.JPEG

Li, X., Liu, H., You, J., Diao, H., Zhao, L., & Wang, W. (2022). Back EVA recycling from c-Si photovoltaic module without damaging solar cell via laser irradiation followed by mechanical peeling. *Waste Management, 137*, 312–318. https://doi.org/10.1016/J.WASMAN.2021.11.024

Li, X., Ma, B., Wang, C., Hu, D., Lü, Y., & Chen, Y. (2023). Recycling and recovery of spent copper—indium—gallium—diselenide (CIGS) solar cells: A review. *International Journal of Minerals, Metallurgy and Materials, 30*(6), 989–1002. https://doi.org/10.1007/S12613-022-2552-Y/METRICS

Lunardi, M. M., Alvarez-Gaitan, J. P., Bilbao, J. I., & Corkish, R. (2018). Comparative life cycle assessment of end-of-life silicon solar photovoltaic modules. *Applied Sciences (Switzerland), 8*(8), 1396. https://doi.org/10.3390/app8081396

Maani, T., Celik, I., Heben, M. J., Ellingson, R. J., & Apul, D. (2020). Environmental impacts of recycling crystalline silicon (c-SI) and cadmium telluride (CDTE) solar panels. *Science of the Total Environment, 735*, 138827. https://doi.org/10.1016/J.SCITOTENV.2020.138827

Mao, D., Yang, S., Ma, L., Ma, W., Yu, Z., Xi, F., et al. (2024). Overview of life cycle assessment of recycling end-of-life photovoltaic panels: A case study of crystalline silicon photovoltaic panels. *Journal of Cleaner Production, 434*, 140320. https://doi.org/10.1016/J.JCLEPRO.2023.140320

Mulazzani, A., Eleftheriadis, P., & Leva, S. (2022). Recycling c-Si PV modules: A review, a proposed energy model and a manufacturing comparison. *Energies, 15*(22), 8419. https://doi.org/10.3390/EN15228419.

Nevala, S. M., Hamuyuni, J., Junnila, T., Sirviö, T., Eisert, S., Wilson, B. P., Serna-Guerrero, R., & Lundström, M. (2019). Electro-hydraulic fragmentation vs conventional crushing of photovoltaic panels–impact on recycling. *Waste Management, 87*, 43–50. https://doi.org/10.1016/J.WASMAN.2019.01.039

NREL. (n.d.). *Cadmium telluride solar cells*. American Institute of Physics Inc. https://doi.org/10.1063/1.4916634.

Preet, S., & Smith, S. T. (2024). A comprehensive review on the recycling technology of silicon based photovoltaic solar panels: Challenges and future outlook. *Journal of Cleaner Production, 448*, 141661. https://doi.org/10.1016/J.JCLEPRO.2024.141661

Rocchetti, L., & Beolchini, F. (2015). Recovery of valuable materials from end-of-life thin-film photovoltaic panels: Environmental impact assessment of different management options. *Journal of Cleaner Production, 89*, 59–64. https://doi.org/10.1016/j.jclepro.2014.11.009

Silveira Camargo, P. S., Petroli, P. A., Andrade de Souza, R., Kerpen, F. S., & Veit, H. M. (2024). CdTe photovoltaic technology: An overview of waste generation, recycling, and raw material demand. *Current Opinion in Green and Sustainable Chemistry, 47*, 100904. https://doi.org/10.1016/j.cogsc.2024.100904

Sinha, P., Cossette, M., & Ménard, J. F. (2012). End-of-life CdTe PV recycling with semiconductor refining. https://www.researchgate.net/publication/308950914_End-of-Life_CdTe_PV_Recycling_with_Semiconductor_Refining.

SolarPower Europe. (2022). EU market outlook for solar power 2022–2026. https://www.solarpowereurope.org/insights/market-outlooks/eu-market-outlook-for-solar-power-2022-2026-2.

Stolz, P., Frischknecht, R., Wambach, K., Sinha, P., & Heath, G. (2017). Life cycle assessment of current photovoltaic module recycling. https://iea-pvps.org/wp-content/uploads/2020/01/Life_Cycle_Assesment_of_Current_Photovoltaic_Module_Recycling_by_Task_12.pdf.

Wambach, K., Heath, G., & Libby, C. (2017). Life cycle inventory of current photovoltaic module recycling processes in Europe. https://iea-pvps.org/wp-content/uploads/2020/01/LCI_of_Current_European_PV_Recycling_WambachHeath_2017_by_Task_12.pdf.

Wambach, K., Libby, S., & Shaw, S. (2024). Advances in photovoltaic module recycling literature review and update to empirical life cycle inventory data and patent review. https://iea-pvps.org/wp-content/uploads/2024/06/IEA-PVPS-T12-28-2024-Report-PV-Recycling-LCI_EPRI.pdf.

Wang, X., Tian, X., Chen, X., Ren, L., & Geng, C. (2022). A review of end-of-life crystalline silicon solar photovoltaic panel recycling technology. *Solar Energy Materials and Solar Cells, 248*, 111976. https://doi.org/10.1016/j.solmat.2022.111976

Wang, J., Feng, Y., & He, Y. (2024). Advancements in recycling technologies for waste CIGS photovoltaic modules. *Nano Energy, 128*, 109847. https://doi.org/10.1016/J.NANOEN.2024.109847

Wang, J., Feng, Y., & He, Y. (2024). The research progress on recycling and resource utilization of waste crystalline silicon photovoltaic modules. *Solar Energy Materials and Solar Cells, 270*, 112804. https://doi.org/10.1016/J.SOLMAT.2024.112804

WEEE Forum. (2021). New WEEE Forum paper: Issues with PV panels & compliance with EPR legislation. https://weee-forum.org/ws_news/weee-forum-releases-pv-paper/.

The manufacturer's authorised representative in the EU is Springer Nature Customer Service Centre GmbH, Europaplatz 3, 69115 Heidelberg, Germany. If you have any concerns regarding our products, please contact ProductSafety@springernature.com

Printed and bound by CPI Group (UK) Ltd, Croydon, CR0 4YY

26/03/2026

02078953-0016